The Mobilities of Ships

As is increasingly recognised within research on mobilities, we live in a world that is ever on the move. Yet studies of mobility have failed to 'go to sea' with the same enthusiasm as mobilities ashore. When we consider mobility, we most often examine those movements that evidently form part of our everyday lives. We forget to look outwards to the sea. Yet ships have played – and continue to play – a significant role in shaping socio-cultural, political and economic life. This book turns our attention to the manifold mobilities that occur at sea through an exploration of the mobilities of ships themselves as well as the movements of objects, subjects and ideas that are mobilised by ships. *The Mobilities of Ships* brings together seven chapters that tack through unexplored waters and move between diverse case studies, including pirate ships, naval vessels and luxury yachts. In so doing, *The Mobilities of Ships* offers a rich insight into the world of shipping mobilities past and present.

This book was published as a special issue of *Mobilities*.

Anyaa Anim-Addo is a Lecturer in Caribbean History at the University of Leeds, UK. She has research interests in the maritime world, the politics of mobility, and race and gender in the post-emancipation Caribbean. She has published research in *Historical Geography, Island Studies* and *Mobilities*.

William Hasty is an Honorary Research Associate at the University of Glasgow. His research explores questions of space, mobility and politics in relation to the seas and seafaring, particularly early-modern piracy.

Kimberley Peters is a Lecturer in Human Geography at Aberystwyth University, UK. Her research focuses on mobility and governance in the context of the sea. She has published work in a number of journals and is the co-editor of the volume *Water Worlds: Human Geographies of the Ocean.*

The Mobilities of Ships

Edited by
Anyaa Anim-Addo, William Hasty and Kimberley Peters

LONDON AND NEW YORK

First published 2015 by Routledge

2 Park Square, Milton Park, Abingdon, Oxon OX14 4RN
711 Third Avenue, New York, NY 10017, USA

Routledge is an imprint of the Taylor & Francis Group, an informa business

First issued in paperback 2017

British Library Cataloguing in Publication Data
A catalogue record for this book is available from the British Library

ISBN 13: 978-1-138-90520-7 (hbk)
ISBN 13: 978-1-138-08276-2 (pbk)

Typeset in Times New Roman
by RefineCatch Limited, Bungay, Suffolk

Publisher's Note
The publisher accepts responsibility for any inconsistencies that may have arisen during the conversion of this book from journal articles to book chapters, namely the possible inclusion of journal terminology.

Contents

Citation Information

The chapters in this book were originally published in *Mobilities*, volume 9, issue 3 (September 2014). When citing this material, please use the original page numbering for each article, as follows:

Chapter 1
The Mobilities of Ships and Shipped Mobilities
Anyaa Anim-Addo, William Hasty & Kimberley Peters
Mobilities, volume 9, issue 3 (September 2014) pp. 337–349

Chapter 2
Metamorphosis Afloat: Pirate Ships, Politics and Process, c.1680–1730
William Hasty
Mobilities, volume 9, issue 3 (September 2014) pp. 350–368

Chapter 3
'The Great Event of the Fortnight': Steamship Rhythms and Colonial Communication
Anyaa Anim-Addo
Mobilities, volume 9, issue 3 (September 2014) pp. 369–383

Chapter 4
Learning 'Large Ideas' Overseas: Discipline, (im)mobility and Political Lives in the Royal Indian Navy Mutiny
Andrew D. Davies
Mobilities, volume 9, issue 3 (September 2014) pp. 384–400

Chapter 5
Unraveling the Politics of Super-rich Mobility: A Study of Crew and Guest on Board Luxury Yachts
Emma Spence
Mobilities, volume 9, issue 3 (September 2014) pp. 401–413

Chapter 6
Tracking (Im)mobilities at Sea: Ships, Boats and Surveillance Strategies
Kimberley Peters
Mobilities, volume 9, issue 3 (September 2014) pp. 414–431

Chapter 7

The Packaging of Efficiency in the Development of the Intermodal Shipping Container
Craig Martin
Mobilities, volume 9, issue 3 (September 2014) pp. 432–451

Please direct any queries you may have about the citations to
clsuk.permissions@cengage.com

Notes on Contributors

Anyaa Anim-Addo is a Lecturer in Caribbean History at the University of Leeds, UK. Her current research focuses on questions of mobility, identity and culture in the post-emancipation Caribbean. Her current project is concerned with labour and leisure in nineteenth-century Caribbean port towns.

Andrew D. Davies is based in the Department of Geography and Planning at the University of Liverpool, UK. He is currently conducting research on transnational activism in late colonial India, and on politics in South Asia more generally.

William Hasty is an Honorary Research Associate at the University of Glasgow. His research explores questions of space, mobility and politics in relation to the seas and seafaring, particularly early-modern piracy.

Craig Martin is a design theorist and cultural geographer based in the School of Design at the University of Edinburgh, UK. His research is concerned with how distributive space can be used to critically interrogate the global contexts of commodity mobilities within the design process, as well as how new forms of material expression associated with 3-D printing technologies, focusing on its impact on what material culture might become.

Kimberley Peters is a Lecturer in Human Geography at Aberystwyth University, UK. Her research focuses on mobility and governance in the context of the sea. She has published work in a number of journals and is the co-editor of the volume *Water Worlds: Human Geographies of the Ocean.*

Emma Spence is a Research Student in the Department of Planning and Geography at Cardiff University, UK. Using an original case study of the luxury super-yacht industry, her research explores the social and cultural geographies of superrich mobility.

INTRODUCTION

The Mobilities of Ships and Shipped Mobilities

ANYAA ANIM-ADDO*, WILLIAM HASTY** & KIMBERLEY PETERS†

**School of History, University of Leeds, West Yorkshire, UK; **School of GeoSciences, University of Edinburgh, Edinburgh, UK; †Department of Geography and Earth Sciences, Aberystwyth University, Ceredigion, UK*

In the burgeoning field of mobilities studies, the seas and all that moves in, on, across and through them, have not been embraced with the same enthusiasm as mobilities ashore. While trains (Verstraete 2002), planes (Adey 2010) and automobiles (Merriman 2007) have received sustained attention, alongside walking subjects (Middleton 2009), wired networks (Graham 2002) and mobile ideas (Law 1986); the ship (a prime figure in seaborne movement) has, for some time, been quietly bobbing in the background (Peters 2010, 1243). It is important to note that the work of the mobilities paradigm has not omitted the politics of sea-based movements entirely (see, for example, recent entries in this journal; Ashmore 2013; Stanley 2008; Straughan and Dixon 2013), but it remains true that mobilities 'at sea' are a vastly underexplored area, with more comprehensive incursions only just beginning to emerge (Anderson and Peters 2014; Birtchnell, Savitzky, and Urry forthcoming; Vannini 2012). This work has helped set in motion a shift towards the seas, following a more general oceanic reorientation within the humanities (see Blum 2010), bringing the rhythms and movements of people, objects, materials, ideas – all manner of things and stuff – into focus through the lens of mobilities thinking.

The following special issue has been inspired by this changing tide, rising off the back of a series of events that have sought to bring the water-world and its manifold maritime mobilities into view. A thumbnail genealogy will illustrate. In 2010, the editors of this collection organised a session at the Royal Geographical Society's annual conference with the Institute of British Geographers (RGS–IBG), entitled 'Geographies of Ships' which sought to explore the spaces, places, times and scales of the ship and the journeys it made possible, in the past and present. This was followed in 2011 by a 'Maritime Roundtable' workshop held collaboratively between Royal Holloway University of London and the University of Glasgow. Here the effort to expand an empirical and conceptual understanding of the mobilities of the ship was extended with the presentation of more specialised and focused papers concerning a range of shipped movements. This was further developed in 2013 with a session explicitly entitled 'Maritime Mobilities' at the RGS–IBG, headed by Emma Spence with a discussant session led by Kimberley Peters. During these workshops and conferences, discussion drifted between notions of ships' mobilities and shipped mobilities; the former being a focus on the movements *of* sea-going vessels and the latter being an interest in what is moved *by* ships in particular circumstances; from the movement of goods, to flows of capital, to the transmission of ideas. The ship as a *moving* thing and *mover* of things, could not easily be separated.

Accordingly, in what follows we introduce this special issue by paying attention to these categories; what we are calling the 'mobilities of ships' and 'shipped mobilities'. We begin by positioning the ship, examining a few instances where studies have attended to notions of mobility in this context, before outlining the long standing omission of ships from the raft of work situated under the rubric of 'mobilities research'. We turn next to the potential of engaging the ship in this field of research, presenting possible avenues of enquiry for future studies of maritime mobilities. Here we focus on the ways in which the mobilities *of* ships and the mobilities facilitated *by* ships may be explored, highlighting where such research furthers the ambitions of mobilities scholarship more generally. We conclude this introduction by surveying the papers that make up this special issue.

Positioning the Ship

Seafaring is an ancient and constant social practice, one that has no discernible beginning or end, one that is thoroughly embedded in the way that humans have understood, explored and lived in the world. The ship is, therefore, one of the oldest technologies of motion (Lavery 2005; Woodman 2005) and as such one of the most important ways that humans have moved. Despite this, however, these mobile and mobilising objects are only now registering on the list of technologies that interest mobilities scholars (Hasty and Peters 2012; Peters 2010). Perhaps this is because in a world of speed, time-space compression and the supposed annihilation of distance (see Massey 1997), they are thought of as 'slow, old fashioned, cumbersome' (Peters 2010, 1245). The world can now be traversed, in a fashion, at the click of the button via technologies that facilitate face-time connections in cyber worlds by tying together disparate spaces. Even physical motion is faster than ever before. The commercialisation of air travel in the 1950s (see Adey 2010) made possible transatlantic journeys in 7 hours as opposed to 7 days. People still move by ship in many different circumstances – with ferries connecting island communities, cruise ships sailing with tourists aboard, migrants crossing the Mediterranean in the dead of night – but with the 'shrinking' of the world, the dominance of the ship has shrunk in our collective imagination. This special issue calls for recognition of its vital role in the movement of things around the world, its centrality to the function of global capitalism, and its importance as a global connector, as well as its role in labour politics, recreational travel and individual experience.

Unsurprisingly, other, faster, more visible methods of moving, by plane, train, car and virtual reality have dominated the examinations of those working in the area of mobilities studies (Adey 2010; Bissell 2010; Merriman 2012). Whilst this work has been of great importance (and continues to have relevance) in understanding the complexities of (post)modern societies, it has, as this introduction argues, been rather terra-centric, focusing largely on landed mobilities and neglecting those that take place at sea. Notably, Adey's work, in particular, has pushed the vision of mobilities upwards, focusing attention on the vertical spaces of motion opened up by air travel (2010, 2013, see also Adey, Whitehead, and Williams 2013). Attending to motions through air space, Adey complicates how we understand the politics of motion through materiality. Accounting for the differences apparent when travelling through air changes how we understand motion on roads or rails, introducing notions of turbulence, for example. Indeed, this work moves us beyond conceptualising a world of surfaces we move on, over and across; to fluid zones we move through (see Forsyth et al. 2013). These conceptual and empirical moves wrought from encounters with vertical mobilities offer useful anchorage points for work on maritime mobilities.

Such a focus will, arguably, becoming increasingly important and, perhaps, necessary in all sorts of ways. The boom in mobility over the past few generations, and our attention to it in research over the past few decades, centres on the (ab)use of carbon-based fuels. As Urry (2012) notes, peak-oil might well coincide with peak-mobility, ushering in a new era of movement, central to which will, undoubtedly, be the ship with its ability to shift across distances great and small under the power of wind, current and muscle. In a recent prospectus on the future of mobilities studies, Cresswell (2014, 4) ruminates briefly on this 'transition', insisting that, when it 'occurs it will necessitate new patterns of movement, new narratives of mobility and new configurations of mobile practice. We will all have to figure out ways of moving differently'. Looking to the seas and ships will certainly be a significant part of this new agenda.

For now, the sea, and ships which travel through it, offer interesting opportunities, akin to those presented by air and airplanes, to move beyond a terrestrial focus to alternate material, elemental spaces that prompt us to rethink mobile action and the relations made through mobilities. What, we might ask, following the work of Palsson (1994), does the movements of sea and ship mean for the act of walking? How does one develop a different way of being in the world through ship-based mobilities? Such opportunities, however, whilst not yet fully apprehended by mobilities scholars, have been gaining increasing interest in other disciplines. Within the field of maritime studies for example, close attention has been paid to the mobile nature of seafaring and oceanic lives (Hasty 2011; Ogborn 2008). Indeed, in this respect, the 'experiences' associated with lives at sea have long been interrogated by scholars of the Royal and merchant navies (see Davies 2013). 'New' maritime history in particular has examined how seafaring and oceanic mobilities were affected by categories of gender, class and race (e.g. see Balachandran 2012; Bolster 1998; Steel 2011). By considering maritime lives on shore as well as at sea, maritime studies have raised questions of (im)mobilities, albeit without using such terms (Creighton and Norling 1996; Land 2007). While maritime historians have much to contribute to the debate, particularly on the experience of oceanic mobilities, such research has yet to commit to full dialogue with mobilities scholarship. The historical contributions to this special issue work towards bridging this divide.

Whilst historians have been mindful of marine and maritime mobilities, so too have geographers and critical theorists, but again largely without engaging the ideas and concepts developed in mobilities studies. Research examining the flow of information, ideas and long distance control through 'actor networks' (see Law 1986), for example, have investigated the potential of the ship as a method by which power was/is mobilised and how this mattered for the diffusion of power from the centre of empires to distant imperial footholds (see also Lambert 2005; Ogborn 2002). Geographers have, moreover, argued that the seas, and the ships that journey upon and through them, open up new spatial imaginaries for mobilising a vision of the world that moves beyond the constructed borders of the nation-state, promising a more fluid way of conceptualising territory and interconnection (Lambert, Martins, and Ogborn 2006). For example, Featherstone's work shows how the ship, as a site of 'dynamic spatial practices', 'dislocates dominant geographical imaginaries' (2008). This work touches upon issues of mobility without really developing a *mobilities* approach to the study of ships or the seas more generally. The papers that feature in this issue variously grapple with the tensions between land and sea, national and international space, fixity and fluidity, through a range of empirical lenses, from the steam packet ship to the luxury yacht to the pirate vessel. The work of this special issue, then, realises some of this promise, bringing mobilities sensibility to bear of the multiple worlds of the ship.

Now is a salient time to rise to this challenge. In recent years, there has been a watery turn across social science and arts and humanities disciplines (Hasty and Peters 2012, 660). This is evident in the range of publications concerning the sea, in disciplines such as History (see Rediker 2007), Architecture (Ryan 2012), English Literature (Sobecki 2008), Art (Mack 2011) and Geography (Anderson and Peters 2014; Steinberg 2001), to name but a few. Amidst this move to unlock the terra-centric and land-locked nature of academic scholarship, the ship, as opposed to the sea itself, remains rather marginal. As Hasty and Peters note in an article that sought to position the ship in geographical study,

> despite marked attention to nautical worlds, the ship, so central to the function of maritime life, remains a largely neglected figure in the literature; a regularly acknowledged but seldom considered feature of the maritime worlds ... (2012, 660)

This 'neglect', as we have noted, is also a feature in the 'mobilities paradigm', a pivotal shift in social sciences which has sought to take seriously the politics of movement, where 'the importance of the systematic movements of people for work and family life, for leisure and pleasure, and for politics and protest' had once been 'ignored and trivialised' (Sheller and Urry 2006, 208). The central argument of this introduction and the papers that follow it is, therefore, that further understanding these phenomena cannot progress without going to sea and aboard the ship.

The advent of mobilities studies has been, in part, a response to 'traditional' transport geographies as a way of understanding what moves, how, where and why (see Shaw and Hesse 2010, 305). Transport geographers have typically started their examinations from an epistemological position based on 'empiricist/positivist assumptions, methods of data collection and modelling' that privilege 'objectivity and truth' over the subjective and multiple meanings and experiences that come with moving (ibid. 2010, 306). Ships have long been part and parcel of a more transport-orientated approach to thinking through movement (see Knowles 2006). Typically, here, attention has been paid to fixed pockets of space, such as the beginning or end of a journey: the ship in port. As Ingold argues, '[t]ransport ... is essentially destination orientated' (2011, 150). The space in-between, movement itself, is relegated, or seen in a simplistic, one-dimensional manner; as a process of 'just moving'. However, as scholars over the past 10 years have shown, movement is never straightforward. The focus on mobilities, an agenda pushed forth by sociologists (Urry 2000), architects (Jensen 2013), geographers (Adey 2010; Cresswell 2006;

Merriman 2012), and others, has sought to unpack the meaning-filled intricacies of movement and challenge a stable, or sedentary, metaphysics (see Cresswell 2006, 27; Sheller and Urry 2006, 208). Ingold's concept of 'wayfaring' develops this also and becomes and important framework for thinking through the significance of ships' mobilities (2011). Ingold too, challenges a sedentarist approach to thinking of the world of experience. He contends that 'lives are not led *inside* places but through, around, to and from them, from and to places elsewhere' (2011, 150). For Ingold, being in the world is not made in containers we typically call 'space' but rather is constituted through movement; lines, or paths that weave together into complex meshworks. This 'logic of inversion' (focusing not on enclosed spaces but the complex inhabitation of world) is useful for thinking of the ship. Often thought of as an enclosed space, the ship; a mobile and still place (externally and internally, respectively); a space that encompasses an interior and exterior, an inside and outside – a space predicated on boundaries – can instead be seen as one that is part of a wider global fabric or *meshwork* of movement; of ties and knots forging places, times and experiences.

Yet it is Cresswell's (2010) recent intervention that has provided, to date, the most comprehensive way in which we may consider the meanings constituted in and through movement. Here he advocates that mobilities are understood 'singularly' – as if the same character or style of movement is shared by all who make it (2010, 17). Instead, Cresswell contends that journeys must be explored in view of the 'force' that drives them, the 'speed' of travel, the 'friction' that stops or prevents the journey, the 'rhythm' that shapes movements, the 'route' of travel, and the 'experiences' of those who live such movements (Cresswell 2010, 17). This agenda suggests useful ways of troubling Euclidean conceptualisations of geometric movement by foregrounding enfolded, multiple, complex, shifting and layered methods, registers and sensations of moving. Accordingly, several papers in this special issue engage with this injunction.

The ubiquity and familiarity of bodies (Bissell 2010; Cresswell 2006; Wylie 2005), mobile technologies (Adey 2010; Merriman 2007; Verstraete 2002), and circulating ideas and information mean that they have quickly come under the theoretical lenses of mobilities scholars embracing this approach. The absence of the sea from explicitly focused mobilities research may, in part, be attributed to the zone in which ships are often put to work: the sea. The sea's position in the background of mobilities studies mirrors the broader tendency in Western thought to see the sea as a lifeless backdrop, a realm distinct and distanced from the land (Steinberg 2013). In being physically beyond the land, the sea has thus been beyond our everyday consciousness. When we now conceive of a world on the move, we have forgotten to look out to the sea, a space Sekula and Birch (2010) argue only enters our vision when 'disaster strikes'. The sea has traditionally been understood in terms of emptiness, disorder and danger, a space antithetical to the land (Connery 2006), though it needs to be noted that such views have their own histories and geographies that are far from universal or constant (see Mack 2011; Raban 1999). As Steinberg (2001) suggests, the sea has for a long time been seen as a space to be crossed as quickly as possible to reach the places that matter, *grounded* centres of capital (Steinberg 2001).

As such, modern and postmodern thinking has conceptualised the sea in a way that seemingly denies from the outset the possibility of seeing it as relevant in the study of mobilities. If the focus of studies of mobility are on the space between points A and B (see Cresswell 2006, 3), but the space between is typically disregarded as a barrier to be traversed for economic and political gain, then we can begin to make sense of why the sea and maritime mobilities have remained obscured for so long, despite the fervour of recent the mobilities turn. Arguably, however, the project of mobilities research, the unpacking of the space between fixed points A and B, the unlocking of the route, rhythm, experience, friction and speed of mobilities (see Cresswell 2010), positively urges us to think about maritime spaces. At sea, there are an abundance of 'gaps' between A's and B's – journeys, moments in transit, lives lived on the move – which have been hitherto overlooked in favour of the apparent fixity and thus importance of points of departure and arrival at either side, on land. This special issue considers the 'gaps' within the broader gap of the sea, namely ships. This issue attends to these overlooked spaces in historical and contemporary mobilities research by focusing on a range of ships (from steam ships to cargo ships), employed in a variety of contexts (from trade to leisure), and the mobilisations that have been made possible by such shipping (in the form of the movement of the things and stuff that make up our material world).

Mobilising the Ship

Though it appears obvious to those whose work is primarily maritime in scope, the case for the ship needs to be made clear: why should we care about ships? Can their omission be read as a sign of their irrelevance to past and present sociocultural, political and economic life? Indeed, alongside the construction of the sea as an empty space, one to be merely crossed to reach places of significance, it has been argued that the sea is irrelevant in any case, because it is not a space of everyday life or 'permanent sedentary habitation' (Steinberg 1999, 386). Most of us do not live our lives at sea, so its bearing on daily existence barely registers. The notion of 'sedentary' is crucial here too. Traditionally, as Cresswell tells us (2006), movement has been deemed to be the antithesis of stable forms of place and senses of belonging, the negative to the positive associations of stillness. The sea – a space of constant flux – has been further relegated therefore, as a space of potential concern. Yet, now that 'issues of mobility are centre stage' (Sheller and Urry 2006, 208), the mobile space between becomes of interest as we seek to unpack 'sociology *beyond* societies' (Urry 2000, emphasis added). The sea is a rich space for the examination of mobilities, being a mobile, four-dimensional plateau in and of itself (Steinberg 2013) that is assembled in wider constellations of motionful elements (winds, the gravitation pull of the moon and so forth – see Jones 2011; Peters 2012). Moreover, it is a space on which, through which, and under which, technologies, people and commodities are made mobile. Furthermore, the sea is also a space through which we may investigate worlds of immobility and the ramifications that result in stoppage within these zones of seemingly incessant flux. As Anim-Addo demonstrates, the shipped mobilities of the Royal Mail Steam Packet Company relied on fixed nodes, such as coaling stations, to make their voyages possible (2011). The range of conditions that immobilise shipping (such as turbulence (Cresswell and Martin 2012)), legal regulations (Peters 2011) and the state of water changing (Vannini and Taggart 2014) are also a key concern in view of understanding economic, political and sociocultural life, in the past and present. Whatever their form and consequences, these (im)mobilities are so often tied up with ships, and as such they are a central concern for this special issue.

Indeed, thus far we have contended that mobilities studies must now pay concerted attention to the ship. However, in what ways should the ship be studied and what potential does it offer for expanding debates and bringing new understandings to the fore in mobilities studies more generally? Firstly, there is need to think about the actual mobility of ships themselves, about why, how and where ships move. Whilst these are seemingly straightforward questions, critically attending to these in light of the mobilities paradigm helps us to unpick the politics tied up with seafaring. For example, investigating how technologies drive the movement of ships can tell us much about the making of the modern world. Anim-Addo's historical work on coaling stations (2011) and the steam ship opens up new understandings of the nodes, such as coaling stations, and flows, such as the movements of ships made possible by coal fuel, that were vital to colonial and extra-imperial movements in the 1800s. Martin's (2013) analysis of shipping container mobilities demonstrates the wider implications of changing rhythms and routines of ships, showing how these processes where connected to shifting ideological and political processes and were ultimately linked to broad social issues such as gentrification in British cities. Following the work of Hodson and Vannini on ferry mobilities between Vancouver Island and the mainland, we might also ask about 'how' ships move in view of theories of time and temporality'(2007, see also Vannini 2012). Vannini (2012) explores how such journeys form an integral part of the life-worlds of islanders, with disruptions, delays and cancellations triggering a host of impacts on service users. Attention to the causes of how ships may become immobilised also draws our attention to the more-than-human elements that can interfere with how ships move and the movement of bodies on ships. Ships may have to deviate from prescribed route ways because of ice, storms, or changing conditions below the surface (see Peters 2012). Ships may also be moved beyond the control of those operating them by elemental forces, resulting in motions of drifting (Peters 2014). Thinking through these questions of how ships move alerts us to the implications that the (im)mobilities of ships have on global supply chains and the impacts of infrastructural breakdown (see Graham and Thrift 2007). Papers in this issue begin to attend to questions of 'how' ships move. Kimberley Peters, for example, considers the role of sea in shaping covert surveillance movements, asking how the materialities of the sea itself prevent the detection of rogue movement on and through oceans. How can such movements be better managed by those conducting surveillance?

The 'where' of shipping movements is a vital concern too. Interrogating the journeys of shipping permits new understandings related to frictions and flows (see Cresswell 2010). Indeed, a study of the 'where' of shipping movements does not need to retreat to an apolitical study of A to B movements, but can be a way to unlock watery politics. Indeed, research in international relations (see Lobo-Guerrero 2011) has explored the role of insurance regimes and shipping. With the threat of piracy still present in many parts of the world, ships often have to change course to avoid 'hot spots' in pinch-points such as channels and straits. In these zones, crews must follow specific regulations on speed, and enact particular body movements onboard to facilitate disciplined observation. Following the movements of ships in terms of questions of 'where' has also gained attention in geography through the 'Waste of the World' project, which sought to trace the movements of ships from life on the ocean to death on the beaches of Chittagong (see Gregson, Crang, and Watkins 2011). This attention to following the 'where' of ships during their lifespan has brought into view the ways in which people, places and times are tied together in previously unrecognised ways. Indeed, the papers in this special issue demonstrate how the ship and sea are vital spaces to contributing to understandings of the land. Anyaa Anim-Addo and Andrew Davies show how the ship and the shore are connected in fresh ways, demonstrating how mobilities at sea are never disconnected from those on land. The movements of ships then, transcend the space of the deck and sea, and allow us to reconfigure understandings of the connected spaces of land and also air. The omission of the ship, as these contributions show, is the omission of a vital frame through which mobilities may be understood. In a world of flows and connections, where space is thought of in relational, fluid terms (Massey 2005) rather than as something fixed, it is entirely appropriate and perhaps necessary to include the seas and ships that sail them as part of this wider global assemblage.

If we are to think about the mobilities *of* ships as important in contributing to a wider picture of fluid, global mobilities, then we must likewise pay attention to the mobilities that are facilitated *by* ships. How, for example, does power disperse, knowledge move, identity shift and goods travel, through the medium of the ship? For example, the ship can become a lens for exploring logistics (see Cowen 2010 and Martin, this issue) and the frictions, turbulence, stasis and flow of supply chains (Sekula and Birch 2010). In short, a host of related mobilities and immobilities can be traced from the ship itself. This project remains to be expanded by mobilities scholars who can trace the politics of mobility through the material and immaterial cargo of ships. Moreover, our attention need not focus only on ships and the mobilities spun from them. Ships can be defined as 'large', 'sea-going' vessels. If the ship allows us to expand upon debates in mobilities studies, in part because of its position at sea, then it figures that our horizons should also expand to encompass other craft at sea too (see Vannini 2009). This issue attempts to demonstrate the potential in moving beyond the ship, to other sea-going vessels, such as the luxury yacht and the small 'tender' boats supplying radio ships in the North Sea. These craft, different in size, design, purpose from larger ocean-going ships present us with an array of mobile narratives from sea that demonstrate once more, the potential of this research area. There is opportunity to extend this scope further still to investigate the range of sea-going craft (and also those that move along rivers and canals), from submarines, to surfers, to rafts, canoes and kayaks, and the mobilities and immobilities that are implicit in their journeys.

Surveying the Articles

The dual themes, of the mobilities of ships and shipped mobilities, which animate this special issue, are aptly opened up for discussion by William Hasty in 'Metamorphosis afloat: the pirate ship as process, c. 1680–1730'. Here, Hasty highlights the mobile creation of spaces at sea, simultaneously underscoring the relation between materialities and the mobilities of ships and identities at sea. Hasty asserts the need to consider the ship as a space that is made and involved in the making of mobilities. In this way, he foregrounds materiality through the making, remaking and contestation of ship-space, positioned as a sort of mobile 'floating assemblage'. The mobility of the (pirate) ship, then, in this account, is not as simple as something which moves (itself and other things) across or through space, but is a thoroughly mutable mobile, an object one the move in every sense.

'"The great event of the fortnight": steamship rhythms and colonial communication' addresses the workings of a nineteenth-century steamship network, focusing in particular on 'friction' within this network. Anyaa Anim-Addo develops the theme of historical immobilities introduced by Hasty, though, unlike him, and also Andrew Davies (in the third paper in this collection), who focus on seafarers afloat, Anim-Addo explores the implications of maritime (im)mobilities ashore. As she demonstrates, such maritime rhythms were negotiated between spaces of sea *and* shore. Thus, while 'Metamorphosis afloat' highlights the inherent mutability of maritime space, "The great event of the fortnight" indicates the changing nature of maritime rhythms, a factor even within individual shipping lines and the practices of particular ships.

Similarly, Andrew Davies examines events in the history of colonialism and considers the circulation of information around the maritime world. Davies shares Anim-Addo's interest in maritime rhythms, but instead undertakes this study from a predominately shipboard or 'at sea' perspective. 'Learning "large ideas" overseas: discipline, routine and political lives in the Royal Indian Navy Mutiny' interrogates seafaring mobilities in a twentieth-century context, and explores race, caste, religion in the formation of identity on naval vessels. Importantly, Davies foregrounds relationships between ship and shore, pointing to Royal Indian Naval vessels as sites of both connection and rupture in relation to land-based norms. In so doing, he presents the ship as a space of colonial ordering and resistance to such ordering processes. Like Hasty, Davies develops an analysis of the ship as assemblage that changes over time and like 'Tracking (im)mobilities at sea', 'Learning "large ideas"' probes questions of discipline and control on the oceans.

'Unravelling the politics of super-rich mobility: a study of crew and guests on board luxury yachts' shifts the perspective from historical subaltern to contemporary elite contexts, demonstrating in the process something of the vast scope that is possible in studies of the ship. Emma Spence probes the significance of 'motive force', 'rhythm' and 'friction' for the super-rich at sea, and examines the interplay between the experiences of the passengers and those of yacht crew, drawing out some of the important politics that necessarily attend the study of mobilities (Cresswell 2010). Although these are very different routines to those discussed by Davies, they nevertheless emerge as multifariously significant, and again offer an insight into life at sea, aboard ships that are spatially ordered and negotiated in direct relation to land-based social hierarchies. Furthermore, Spence develops Davies's attention to the formation of identities at sea by foregrounding the some of the ways that the super-rich perform these mobile maritime identities.

'Tracking (Im)mobilities at Sea: Ships, Boats and Surveillance Strategies' directly engage the theme of maritime power relations introduced by Hasty and Davies, but shifts the focus towards formal state attempts to exercise control over and through ships at sea. Kimberley Peters asserts that the materiality of the ocean poses particular challenges for surveillance and, in this way, she reflects on what happens when land-based challenges are translated off-shore. The call of Hasty to take seriously the mobile materiality of the ship, is mirrored by Peters' injunction to interrogate the changing materiality of the ocean through which the ship moves. For both ship and sea, materiality and its inherent mobility when in contact with elemental and social forces matters a great deal. Hasty indicates that the material form of the ship was far from fixed, and Peters foregrounds not only the implications of the ambiguous legal status of vessels in international waters, but also the effect of the materiality of the ocean on attempts to police boundaries at sea.

'The packaging of efficiency in the development of the intermodal shipping container' interrogates shipped mobilities through a focus on the standardisation of the packaging of goods moved at sea. Craig Martin argues for a greater appreciation of the socio-economic significance of containerisation in the development of port cities and maritime communities, and highlights the importance of container shipping in globalisation and contemporary commerce. In so doing, he presents an analysis of maritime infrastructure, which echoes some of the concerns of Anim-Addo about the importance of the rhythms and routines developed in service of trade from a different theoretical vantage point. As for both Hasty and Davies, the concept of 'assemblage' proves fruitful for Martin, though in this case it is not to consider the ship itself, but rather in the broader set of maritime logistics into which the ship is bound. In this paper, Martin demonstrates the social implications of the desire to engineer ever greater mobile efficiency through the material form of shipping and maritime infrastructure.

As well as making important empirical, conceptual and methodological contributions to the study of ships, the maritime world, mobilities and much else besides, the articles in this special issue suggest the great potential for studies of maritime matters, including, but not limited to ships, through a mobilities lens. The papers, each in different ways, highlight the emergent, instable and slippery nature of the mobilities of ships and shipped mobilities. The ship, the lives of those who sail them and maritime logistics are theorised and explicated in a multitude of ways, notably as a mobile and mobilising assemblage (see papers by Hasty, Davies and Martin). The importance of ever-fluctuating rhythms and routines to the operation of ships and the wider world into which they fit also emerges in this special issue (see Anim-Addo's contribution). It is also made clear that attempts to assert order and control afloat exist in constant tension with the changeable nature of the maritime world (see Peters), and that social practices performed at sea are negotiated between actors who are influenced by the ship (contribution by Spence). By taking consideration of mobilities offshore, and by examining the mobility of the ship and shipped mobilities in historical and contemporary contexts, this special issue develops, extends and enlivens the emerging concern with space, mobility and materiality at sea.

Acknowledgements

We would like to thank the participants of the many conference sessions and workshops from which this special issue emerged, and the anonymous reviewers for their comments on earlier drafts of this introduction. Our special thanks goes to Pennie Drinkall for her patience and assistance in helping us to complete the project.

References

Adey, P. 2010. *Aerial Life: Spaces, Mobilities, Affects*. Oxford: Wiley-Blackwell.

Adey, P. 2013. *Air*. London: Reaktion.

Adey, P., M. Whitehead, and A. Williams, eds. 2013. *From above: War, Violence and Verticality*. London: Hurst & Co.

Anderson, J., and K. Peters, eds. 2014. *Water Worlds: Human Geographies of the Ocean*. Farnham: Ashgate.

Anim-Addo, A. 2011. "'A Wretched and Slave-like Mode of Labor': Slavery, Emancipation and the Royal Mail Steam Packet Company's Coaling Stations." *Historical Geography* 39: 65–84.

Ashmore, J. 2013. "Slowing down Mobilities: Passengering on an Inter-war Ocean Liner." *Mobilities* 8: 595–611.

Balachandran, G. 2012. *Globalizing Labour: Indian Seafarers and World Shipping c. 1870–1945*. New Delhi: Oxford University Press.

Birtchnell, T., S. Savitzky, and J. Urry. forthcoming. *Cargomobilities: Moving Materials in a Global Age*. London: Routledge.

Bissell, D. 2010. "Passenger Mobilities: Affective Atmospheres and the Sociality of Public Transport." *Environment and Planning D: Society and Space* 28: 270–289.

Blum, H. 2010. "The Prospect of Oceanic Studies." *Publications of the Modern Languages Association of America* 125 (3): 670–677.

Bolster, W. 1998. *Black Jacks: African American Seamen in the Age of Sail*. Cambridge, MA: Harvard University Press.

Connery, C. 2006. "There was No More Sea: The Supersession of the Ocean, from the Bible to Cyber-space." *Journal of Historical Geography* 32 (3): 494–511.

Cowen, D. 2010. "A Geography of Logistics: Market Authority and the Security of Supply Chains." *Annals of the Association of American Geographers* 100 (3): 600–620.

Creighton, M., and L. Norling. 1996. *Iron Men, Wooden Women: Gender and Seafaring in the Atlantic World, 1700–1920*. Baltimore, MD: The John Hopkins University Press.

Cresswell, T. 2006. *On the Move: Mobility in the Modern Western World*. London: Routledge.

Cresswell, T. 2010. "Towards a Politics of Mobility." *Environment and Planning D: Society and Space* 28 (1): 17–31.

Cresswell, T. 2014. "Mobilities III: Moving on." *Progress in Human Geography*. (Early Online Publication). doi:10.1177/0309132514530316.

Cresswell, T., and C. Martin. 2012. "On Turbulence: Entanglements of Disorder and Order on a Devon Beach." *Tijdschrift Voor Economische En Sociale Geografie* 103: 516–529.

Davies, A. 2013. "From 'Landsman' to 'Seaman'? Colonial Discipline, Organisation and Resistance in the Royal Indian Navy, 1946." *Social and Cultural Geography* 14: 868–887.

Featherstone, D. 2008. *Resistance, Space and Political Identities: The Making of Counter-global Networks*. Oxford: Wiley-Blackwell.

Forsyth, I., H. Lorimer, P. Merriman, and J. Robinson. 2013. "What are Surfaces?" *Environment and Planning A* 45: 1013–1020.

Graham, S. 2002. "Constructing Premium Network Spaces: Reflections on Infrastructure Networks and Contemporary Urban Development." *International Journal of Urban and Regional Research* 24: 183–200.

Graham, S., and N. Thrift. 2007. "Out of Order: Understanding Repair and Maintenance." *Theory, Culture and Society* 24: 1–25.

Gregson, N., and M. Crang, and H. Watkins. 2011. "Souvenir Salvage and the Death of Great Naval Ships." *Journal of Material Culture* 16: 301–324.

Hasty, W. 2011. "Piracy and the Production of Knowledge in the Travels of William Dampier, c.1679–1688." *Journal of Historical Geography* 37 (1): 40–54.

Hasty, W., and K. Peters. 2012. "The Ship in Geography and the Geographies of Ships." *Geography Compass* 6 (11): 660–676.

Hodson, J., and P. Vannini. 2007. "Island Time: The Media Logic and Ritual of Ferry Commuting on Gabriola Island, BC." *Canadian Journal of Communication* 32: 261–275.

Ingold, T. 2011. *Being Alive: Essays on Movement, Knowledge and Description*. Abingdon: Routledge.

Jensen, O. B. 2013. *Staging Mobilities*. London: Routledge.

Jones, O. 2011. "Lunar–solar Rhythmpatterns: Towards the Material Cultures of Tides." *Environment and Planning A* 43: 2285–2303.

Knowles, R. D. 2006. "Transport Shaping Space: Differential Collapse in Time-space." *Journal of Transport Geography* 14: 407–425.

Lambert, D. 2005. "The Counter-revolutionary Atlantic: White West Indian Petitions and Proslavery Networks." *Social and Cultural Geography* 6 (3): 405–420.

Lambert, D., L. Martins, and M. Ogborn. 2006. "Currents, Visions and Voyages: Historical Geographies of the Sea." *Journal of Historical Geography* 32: 479–493.

Land, I. 2007. "Tidal Waves: The New Coastal History." *Journal of Social History* 40 (3): 731–743.

Lavery, B. 2005. *Ship: 5000 Years of Maritime Adventure*. London: Dorling Kindersley.

Law, J. 1986. "On Methods of Long-distance Control: Vessels, Navigation and the Portuguese Route to India." In *Power, Action and Belief: A New Sociology of Knowledge?* edited by J. Law, 234–263. London: Routledge.
Lobo-Guerrero, L. 2011. *Insuring Security: Biopolitics, Security and Risk*. London: Routledge.
Mack, J. 2011. *The Sea: A Cultural History*. London: Reaktion.
Martin, C. 2013. "Shipping Container Mobilities, Seamless Compatibility, and the Global Surface of Logistical Integration." *Environment and Planning A* 45: 1021–1036.
Massey, D. 1997. "A Global Sense of Place." In *Reading Human Geography: The Poetics and Politics of Enquiry*, edited by D. Gregory and T. Barnes, 315–323. London: Arnold.
Massey, D. 2005. *For Space*. London: Sage.
Merriman, P. 2007. *Driving Spaces: A Cultural-historical Geography of England's M1 Motorway*. Oxford: Wiley-Blackwell.
Merriman, P. 2012. *Mobility, Space, and Culture*. London: Routledge.
Middleton, J. 2009. "'Stepping in Time': Walking, Time, and Space in the City." *Environment and Planning A* 41 (8): 1943–1961.
Ogborn, M. 2002. "Writing Travels: Power, Knowledge and Ritual on the English East India Company's Early Voyages." *Transactions of the Institute of British Geographers* 27 (2): 155–171.
Ogborn, M. 2008. *Global Lives: Britain and the World 1550–1800*. Cambridge: Cambridge University Press.
Palsson, G. 1994. "Enskillment at Sea." *Man* 29 (1): 875–900.
Peters, K. 2010. "Future Promises for Contemporary Social and Cultural Geographies of the Sea." *Geography Compass* 4 (9): 1260–1272.
Peters, K. 2011. "Sinking the Radio Pirates: Exploring British Strategies of Governance in the North Sea, 1964–1991." *Area* 43 (3): 281–287.
Peters, K. 2012. "Manipulating Material Hydro-worlds: Rethinking Human and More-than-human Relationality through Offshore Radio Piracy." *Environment and Planning A* 44 (5): 1241–1254.
Peters, K. 2014. "Taking More-than-human Geographies to Sea: Ocean Natures and Offshore Radio Piracy." In *Waterworlds: Human Geographies of the Ocean*, edited by J. Anderson and K. Peters, 177–191. Farnham: Ashgate.
Raban, J. 1999. *Passage to Juneau: A Sea and Its Meanings*. Basingstoke: Picador.
Rediker, M. 2007. *The Slave Ship: A Human History*. London: John Murray.
Ryan, A. 2012. *Where Land Meets Sea: Coastal Explorations of Landscape, Representation and Spatial Experience*. Farnham: Ashgate.
Sekula, A., and N. Birch. 2010. *The Forgotten Space*. Amsterdam: Doc.Eye Film.
Shaw, J., and M. Hesse. 2010. "Transport, Geography and the 'New' Mobilities." *Transactions of the Institute of British Geographers* 35: 305–312.
Sheller, M., and J. Urry. 2006. "The New Mobilities Paradigm." *Environment and Planning A* 38: 207–226.
Sobecki, S. 2008. *The Sea and Medieval English Literature*. Cambridge: D S Brewer.
Stanley, J. 2008. "Co-venturing Consumers 'Travel Back': Ships' Stewardesses and Their Female Passengers, 1919–551." *Mobilities* 3: 437–454.
Steel, F. 2011. *Oceania under Steam: Sea Transport and the Cultures of Colonialism, c. 1870–1914*. Manchester: Manchester University Press.
Steinberg, P. E. 1999. "Navigating to Multiple Horizons: Towards a Geography of Ocean Space." *Professional Geographer* 51 (3): 366–375.
Steinberg, P. E. 2001. *The Social Construction of the Ocean*. Cambridge: Cambridge University Press.
Steinberg, P. E. 2013. "Of Other Seas: Metaphors and Materialities in Maritime Regions." *Atlantic Studies* 10 (2): 156–169.
Straughan, E. and D. Dixon 2013. "Rhythm and Mobility in the Inner and Outer Hebrides: Archipelago as Art-science Research Site." *Mobilities*. (Early Online Publication). doi:10.1080/17450101.2013.844926.
Urry, J. 2000. *Sociology beyond Societies*. London: Routledge.
Urry, J. 2012. "Do Mobile Lives Have a Future?" *Tijdschrift Voor Economische En Sociale Geografie* 103: 566–576.
Vannini, P., ed. 2009. *The Cultures of Alternative Mobilities: Routes Less Travelled*. Farnham: Ashgate.
Vannini, P. 2012. *Ferry Tales: Mobility, Place and Time on Canada's West Coast*. London: Routledge.

Vannini, P., and J. Taggart. 2014. "The Day We Drove the Ice Road (and Lived to Tell the Tale about It) of Deltas, Ice Roads, Waterscapes and Other Meshworks." In *Waterworlds: Human Geographies of the Ocean*, edited by J. Anderson and K. Peters, 89–102. Farnham: Ashgate.

Verstraete, G. 2002. "Railroading America: Towards a Material Study of the Nation." *Theory Culture and Society* 19: 145–159.

Woodman, R. 2005. *The History of the Ship: The Comprehensive Story of Seafaring from the Earliest times to the Present Day*. London: Conway Maritime Press.

Wylie, J. 2005. "A Single Day's Walking: Narrating Self and Landscape on the South West Coast Path." *Transactions of the Institute of British Geographers* 30: 234–247.

Metamorphosis Afloat: Pirate Ships, Politics and Process, c.1680–1730

WILLIAM HASTY

School of Geosciences, University of Edinburgh, Edinburgh, UK

ABSTRACT *This paper follows some late-seventeenth and early eighteenth century pirate ships, focusing upon the moments when these most enigmatic and elusive of ocean-going vessels were appropriated and inhabited by mutinous mariners who literally risked their necks to take charge of them. This paper builds upon recent work in mobilities and oceanic studies which is developing more materialist perspectives as a means for better understanding the seas and ships as lived, dynamic spaces. By exploring some of the ways that pirate ships were crafted and modified, and then occupied, at the turn of the eighteenth century, this paper contributes new perspectives on the formation of piratical spaces and identities, and in the process, the role of mobilities and spatialities in creating spaces afloat. The paper argues for a greater acknowledgement of the role of process in the making of space and mobility at sea as a means of better understanding the complex geographies of the pirate ship and the experiences of those who sailed aboard them.*

Introduction

This paper follows some late-seventeenth and early eighteenth century pirate ships, focusing upon the moments when these most enigmatic and elusive of ocean-going vessels were appropriated and inhabited by mutinous mariners who literally risked their necks to take charge of them. The most expensive and technologically advanced objects in existence (Adams 2001), the ships seized by pirates were their homes, where their founding moment of rebellion and subsequent life *as* pirates unfolded, where they lived in contempt of international efforts towards their extirpation (Earle 2004). However, one theorises pirates, whether as opportunistic thieves (Earle 2004), rational choice-making entrepreneurs (Leeson 2009), savvy lawyers at the edges of order (Benton 2010) or politically radical rebels (Rediker 2004), in each and every case, the practices of the pirate were inescapably entangled with the geographies of the ship. Thus far, in accounts of the pirate ship, neither spatiality nor mobility has been particularly prominent or well-developed analytical levers. The potential for picking through histories of piracy and the ship more generally are therefore significant, since, as Merriman et al. (2012, 7) insist, thinking with space (and mobility, it might well be added) is a sure means of unlocking narratives too quickly closed down otherwise: 'spatiality can disrupt theories that have not taken it seriously'.

In drawing attention to the material practices involved in the transition of a ship to a pirate ship, this paper presents the ship as a space defined by process, as a site wherein form and function were subject to continual negotiation, re-imagined and reshaped by social, political and practical imperatives. Piracy, then, by extension, can be thought of as a process or set of processes, as an inherently relational practice, rather than one which is in any way fixed or stable. Piracy in this period was always the result of radical shipboard transformations, the changing social and material shape of the ship; a sort of metamorphosis afloat.[1] The ship became the pirate ship as it changed from one state to another, not in a pre-given sense (in the way that metamorphosis might describe the transformation of one stage of a species into another, as in the case of the caterpillar into the butterfly), but in a fluid way, through a contested set of transformational practices. Pirates seized space at sea and made it their own, enacting their own politics, social arrangements and cultural practices through ship-space and the modifications they made to it.

This approach speaks to concerns with mobilities in important ways. The ship, Cresswell (2011, 555) argues, is a good place to re-think and re-tool the 'new mobilities paradigm' (Sheller and Urry 2006). Firstly, ships seem to present a means of raising 'an awareness of the mobilities of the past', encouraging a stronger, deeper historical consciousness in the work of the new mobilities paradigm (Cresswell 2010, 28). Secondly, venturing into earth's watery realms and engaging the ship seems to promise a vast range of new empirical, methodological and conceptual challenges for mobilities scholars (Cresswell 2011, 555; Hasty and Peters 2012), perhaps encouraging an encounter with what Chambers (2010, 3) has recently called 'maritime criticism': a shift in thinking which 'sets existing knowledge afloat: not to drown or cancel it, but rather to expose it to unsuspected questions and unauthorised interruptions'. Taking mobilities to sea aboard the ship, then, allows us to see things previously viewed as 'bounded and fixed, stable and permanent, in terms of flows and fluids – in terms of movement' (Adey 2006, 77).

In its attention to early modern pirate ships, this paper embraces these challenges. Moreover, the pirate ship presents a serious challenge for a problematic assumption about the potential mobilities of the ship embedded in Cresswell's (2011, 555) 'watery mobilities' injunction; namely, the idea that ships are, at any given point, 'essentially the same thing as they have always been, give or take a few modifications in power supply and navigation'. Following Blum's (2010, 669) assertion that the 'sea is not a metaphor' and Steinberg's (2013, 157) insistence that we 'directly engage the ocean's fluid mobility and its tactile materiality', this paper contends that accounts of the ship must also venture beyond the comforts of the metaphor – the ship as 'heterotopia' (Foucault 1986), the ship as 'natural contract' (Serres 1995), to name but two – and embrace what Blum (2010, 670) calls 'the material conditions and praxis of the maritime world' (see also Peters 2012). Indeed, to better understand the important socio-spatial relations that have for centuries been woven around ships, the people who sailed them and the locations they visited and inhabited at different times, we must precisely avoid thinking of ships in this static way, and instead look for the ways that ships were (and are) inherently physical and dynamic, lived and contested sites of multiple mobilities. We must see the ship not only as a mobilising force – a technology for moving things across space – but as a site itself inherently constituted by ideas and practices of movement.

To examine an act of piracy, then, is to see a ship shifting in states – material, social and political – and, as a result, to see space itself in flux. The remainder of this paper proceeds by first considering the conceptual underpinnings of a mobilities analysis of pirate ships, and then presenting an account of the pirate ship as a site of material transformation and socio-political contestation. This focus follows Massey (2005, 11) in thinking about space (and mobilities) as 'always in process … never a closed system'. As Steinberg has recently argued, thinking this way, about fluidities, enables new perspectives on 'space itself and how it is produced (and reproduces itself) within the dynamics of spatial assemblages' (Steinberg 2013, 163). This paper attempts to realise some of this promise and seeks to develop recent work in geography and mobilities studies more generally which has, as Adey (2006, 90) suggests, re-imagined space with movement and process at the heart of things: 'To be sure, process rules. Space is never still, it can never just be – because mobilities compose material process and becomings. They constitute new apprehensions of space'. It is arguably through attending to these mobile and mobilising processes that we can better understand the politics of piratical identities, the social and cultural worlds of seafarers and the nature of being afloat, in the past, more generally.

Ships, Mobilities and Pirates

In *The Black Atlantic*, Gilroy (1993, 4) famously positions the ship as 'a living, micro-cultural, micro-political system in motion'. As Steinberg (2013, 158) has recently joked, however, delving 'into Gilroy's Black Atlantic, one never gets wet'. The elision of the sea as a physical entity in the influential work of Gilroy and many others (e.g. Virilio 1977) is mirrored in a similar neglect of ships' materialities; their lived and lively qualities overlooked as they serve as metaphors in the tales of connection and flow their movements inspired. Take, for example, Casarino's (2002, 19) claim that '[t]he ship never travels, never goes anywhere, never even moves'. Here, the ship is not really a ship; it is a cypher for something else, a metaphor. In essence, the problem with the ship as metaphor, and only metaphor, is that it must remain fixed and pre-given to successfully occupy that role. Smith and Katz (1993, 68) have argued this in relation to spatial metaphors more generally: '[m]etaphor works by invoking one meaning system to explain or clarify another. The first meaning system is apparently concrete, well understood, unproblematic, and evokes the familiar'. My purpose here is not to dismiss the metaphor entirely, rather, it is to encourage a move beyond the kinds of metaphorical formulations of the ship which ignore the vibrancy inherent to notions of spatiality and mobility.[2] For this reason, while the ship never moved beyond the realms of metaphor for Gilroy, his injunction to others to see ships as living, dynamic spaces remains salient; indeed, despite the innumerable accounts of famous, infamous and supposedly archetypal ships that litter scholarly and popular histories, 'the ship as a physical entity has barely been explored' (Jarvis 2007, 52).

New geographies of the ship are steadily emerging and this situation is beginning to change (Hasty and Peters 2012). Recent work on, amongst other things, ocean liners (Ashmore 2013), Royal Mail steamers (Anim-Addo 2011), pirate ships (Hasty

2011), passenger ferries (Vannini 2011), educational cruisers (della Dora 2010), private yachts (Ryan 2006) and a commuter boat (Boshier 2009) has taken significant steps towards explicating the complex socio-spatial arrangements to be found afloat at various times and places. What this work does well is that it acknowledges the diversity of material and social structures which constitute a given vessel and its movements to better account for the formation of particular identities, social relations, ideas and knowledges, senses of place and so on. Ashmore (2013, 14), for example, in considering the 'processuality of passengering' aboard steamships in the early twentieth century, suggests that '[b]eing mobile over a long period spent at sea is … generative of specific affects and feelings in the humans who dwell in and travel through these watery spaces'. In particular, he points out that an 'important element shaping the experience of travel is the physical form of the modality travelled in. The physical layout of the ocean liner as a mobile form is very different to that of other modalities that are often the focus of study' (Ashmore 2013, 8). In short, this emerging body of exemplary work situates the ship as a material entity and its multiple mobilities in the foreground of the story in the telling. With few exceptions, however, this attention has thus far been directed almost exclusively at what might be called 'ship-shaped relations', those interactions induced and produced by being aboard ship (Hasty and Peters 2012, 664). For example, for Ashmore (2013, 15), 'the ship is a more or less durable structure', while 'the affects that work through it are more transitory'. As a result of this kind of thinking, another important facet of this socio-spatial dialectic remains relatively untouched; namely, the role of socio-political relations in shaping the spatialities and mobilities afloat. To consider this is to ask to what extent the form and function of the ship might reflect the on-going social and political exchanges both aboard and in other locations.

Though there are many ways to divide the literature on ships – by discipline, time period, technical type, method of power and so on – a significant and influential section can be separated into two broad streams according to their position vis-à-vis shipboard social relations. On one side, we have those who see the ship as a clear extension of the norms, values and structures of landed society and on the other, we have those who find exactly the opposite, profound difference and otherness. For the former, Conrad's observation that the ship was 'fragment detached from the earth' which 'went on lonely and swift like a small planet' rings true (cited in Casarino 2002, 19). For example, Pearson (2010, 9–10) insists that the social order of the ship in the early modern period replicated 'quite precisely landed society, as seen in authority structures, food, reaction to stress, the comforts of religion and so on'. Conversely, for the latter, the ship is actually where we find the structures of landed power contested and subverted, 'a place to which and in which the ideas and practices of revolutionaries … escaped, re-formed, circulated, and persisted' (Linebaugh and Rediker 2000, 144–145). Indeed, rather than finding a fragment of the earth detached, the early modern ship – especially the pirate ship – was a site in which one could find 'the world turned upside down' (Rediker 2004, 61; after Hill 1972). While acknowledging the clear tensions between these positions, there are a couple of important, and somewhat problematic, assumptions they hold in common. In the first instance, both helpfully envision the ship in broadly relational terms, as formed and maintained through social, political and cultural bonds, held in relation to other spaces, either those of dominant polities (Pearson 2010) or pan-oceanic subaltern networks (Linebaugh and Rediker 2000). Secondly, within this, we can see that spatiality *is* acknowledged, and that both follow Dening (1992, 19) in his pithy observation that '[s]pace and the language used to describe it make a ship'. They acknowledge that (most) ships were riven with spatially engrained hierarchies, what Jarvis (2007, 61) calls the 'complex social geography' of the ship'.

Space is theorised in these accounts in rather problematic ways, oscillating between fixity and relationality, and without an acknowledgement of the centrality of process to space (Massey 2005). As Featherstone (2005, 392–393) has noted of Linebaugh and Rediker (2000) in particular, there is a tendency to 'treat space as a fixed backdrop to political activity. Ideas, tactics and radical experiences flow and move across space, but these circuits remain unchanged through these processes'. A similar critique could be made about the way mobility is understood in this literature. While movements, flows and connections are rightly foregrounded in much of the work on maritime history, mobilities are too often conceived of in too simplistic a fashion. The ship is usefully understood as a mobilising force, as the thing that moves people, ideas and other things around, as the 'engine of commerce, the machine of empire', and the primary 'means of communication between continents' (Linebaugh and Rediker 2000, 150–152). More than this, however, the movement of the ship was central to the seafaring way of life: 'Mobility, fluidity, and dispersion were intrinsic to the seaman's life … Seamen were in many ways nomads, and their mobility ensured a rapid diffusion of their culture' (Rediker 1987, 159). Notwithstanding the important contributions made by this literature to our understanding of the cross-oceanic connections facilitated by the movement of ships, these analyses of seagoing movements stop short of fully engaging mobilities theory. This is true of work on the sea more generally. On the possibility of directly engaging the 'very geophysical mobility' of the sea in our accounts of mobilities, Steinberg (2013, 165) writes:

> From this perspective, the ocean becomes the object of our focus not because it is a space that *facilitates* movement – the space across which things move – but because it is a space that is *constituted by* and *constitutive of* movement.

The ship too is a space that deserves attention not only on the basis of its ability to move things, but because it is a space clearly created, inhabited and understood in various ways according to differently experienced mobilities and immobilities. It is a space constituted by and constitutive of movement. The ship, and in particular the pirate ship, appears as a floating assemblage on the move, carrying things across space, but also shifting form in space: transmuting as well as transporting. Such a focus speaks to the very core of mobilities theory: 'Mobilities cannot be described without attention to the necessary spatial infrastructure and institutional moorings that configure and enable mobilities' (Hannam, Sheller, and Urry 2006, 3).

This paper builds on and develops work on the pirate ship and the mobilities of ships by foregrounding the processual nature of the pirate ship as a space. These spaces are notoriously elusive for the historian: 'Of the life on board buccaneer and pirate ships only a somewhat hazy and incomplete picture reaches us' (Gosse 1924, 21). Perhaps, as a result of this perceived shortage of reliable or recognisable evidence, the pirate ship is much mythologised, and now casts an instantly identifiable shadow. A number of academic accounts have done much to cast new light upon the pirate ship, bringing into focus the alternative social structures enacted aboard (Rediker 2004), the complex cross-cultural interactions it engendered (Bialuschewski 2008), the violent practices it witnessed (Earle 2004) and the ingenious legal posturing it promoted (Benton 2010). Through these studies, our understanding of the pirate ship moves beyond simple myth and romance, but, by and large, they fail to directly engage the pirate ship as a mobile, physical entity. The ship *is* presented as somewhat dynamic, but *only* in the moment that the pirates seize it and take control – otherwise it is fixed. After the ship becomes the pirates ship (with the exception of Benton's (2010) work on the pirate ship as a space of alternative legal practices) space stands still, held in place by a set of assumptions about the intersections of identity, practice, space and mobilities.

Drawing on the insights of mobilities scholars and geographers, it is possible to get beyond the static space of the pirate ship presented in the literature and develop an account of the ship which embraces the processual nature of ship-space. As Ashmore (2013, 15) argues, there is 'a necessity to consider the materialities and socialities specific to different modalities of mobility', focusing not on the 'static' but on those which are 'continually emergent' as they 'cohere, disperse and reappear in new assemblages'. This attention to the assemblage of physical *and* social things chimes with another strand of recent work in mobilities. As Strohmayer (2011, 121) argues of the bridge, it 'is not a unified, homogenous structure. Nor do bridges function in a unilateral manner. As a result, the particular forms of mobility facilitated and enabled by their presence are historically specific and anchored in concrete existing contexts'. A similar sensibility could offer new ways of interrogating the pirate ship, by working through the multiple mobilites enacted in and constitutive of the material structures of the ship and the social, political and cultural interactions engendered therein.

Metamorphosis Afloat

Stillness is a rare thing. The sea, or 'hydroworld' as Peters (2012, 1242) puts it, is characterised by 'instability, and sheer, motionful, dynamic power'. Even in the dreaded doldrums, ships and all they contained were subject to the motions of the sea. The ship not only moves between points on a plane, but in other directions and dimensions too, with the swell of the waves. Anchors don't hold ships still, they hold them in place, and only to an extent. These forces are constant, and as they moved the ship they took their toll on its fabric and form. The ships timbers were usually dressed in sticky tar to preserve them against the corrosive power of saltwater. Ingress was inevitable, slowly but surely the wooden ship would succumb to mould and rot. Other organisms, like the Toledo worm, clung, gnawed and burrowed their way through the wood, eating the ship and altering its structural properties as they went. Thundering waves, howling winds, unknown rocks, even the odd piece of flotsam or jetsam could substantially alter the fabric of the ship, or, at least, call on the ship's hands to make alterations. The ship, as absolutely any sailor in the age of sail would have been able to testify, was a living, breathing entity. Far from barely changing over the course of centuries, the ship was a space literally on the move, on the shift, from one moment to the next, in subtle and profound, intentional and inevitable ways. It was never static and it was never stable. It was a body decaying from its launch, kept afloat through the continuous labours of those who sailed it, modifying as they went to keep it moving. This speaks to a key contention emerging in the study of mobilities and materialities:

> … in reality the body is changing form at every moment; or rather, there is no form, since form is immobile and the reality is movement. What is real is the continual change of form: form is only a snapshot view of a transition. (Bergson, quoted in Adey 2006, 77)

As well as the changes necessitated by the force of the elements, pirates were apparently inclined to alter their vessels according to another set of forces, those emanating from political and social imaginaries. Central to this paper is the argument that pirates took ships and shaped them, socially, politically *and* materially to suit their own ends, that, in other words, the pirate ship was a mobile and mobilising force at sea but it was also an inherently mutable entity. Not only was the ship not merely a metaphor, but it was a living thing, a transformative, material site of contested politics. In what now follows, the nature and extent of these changes are considered in three important ways, looking first to concerns around speed and practicality, before moving on to examine the relations between materiality and politics on the decks of pirate ships, before ending with a discussion of the social and political implications of this mutability for our understanding of the pirate ship and those we find aboard them.

Moving Fast and Getting Close

On 1 April 1719, pirates boarded the slave ship *Bird Galley* at the mouth of the Sierra Leone River and took its captain captive. As was the custom on such occasions, William Snelgrave found himself in conference with his captors, who wanted to know how well his ship sailed, 'both large, and on a wind' (Snelgrave 1734, 213). Pirates' ships were never built for them and their particular purposes. They worked with what they could get, crafting space afloat, on-the-fly, through assiduous and resourceful re-appropriation. First and foremost, they sought vessels which appeared seaworthy. They also preferred ships that were well-armed (or displayed the potential to bear significant arms) and those that were fast. His reply that his ship, a galley-built vessel designed for speed, sailed 'very well' seemed to confirm the pirates' own favourable impressions, and led to their commander throwing his hat in the air and declaring of the *Bird Galley* that 'She would make a fine Pirate Man of War' (Snelgrave 1734, 213).

Mobility, in the most straightforward sense, was an essential feature of piracy, indeed in his *History of the Pyrates* – arguably the most important and heavily debated source for piratical histories of the so-called golden age (c.1680–1730) – Johnson (1726, 168) opines that 'a light pair of heels was of great use either to take or to escape being taken' at sea. As well as taking the opportunity to seize notably fast ships, we can find evidence of pirates' deep understanding of the kinds of modification and maintenance required to ensure speed, nothing less than a matter of life and death for them. Having spent a month in their company, Snelgrave serves as a good witness to the moment when a ship became a pirate ship. One thing he records about this event is that as soon as the pirates had decided that his ship should be theirs, 'all hands went to work to clear the Ship, by throwing over board Bales of Woollen Goods; Cases of India Goods; with many other things of great Value' (Snelgrave 1734, 223). This kind of thing has often been interpreted as either the mark of wanton irrationality (e.g. Earle 2004) or anti-capitalist radicalism (e.g. Rediker, 2004), though it more likely speaks of the pirates' desire to lighten the load of the ship, thereby rendering it clearer in case of engagement and more agile afloat. The ship and its constituent parts were never sacred in the hands of pirates. The ship was their home at sea, but it was also a tool for transport, one that had to be honed for the purpose of the pirate through modification.

We can look to the practice of careening, extremely common among pirates during the late-seventeenth and early eighteenth centuries, to see this awareness of mobility and speed more clearly still. Careening a large wooden ship was an arduous task, one involving significant labour starting with the hauling ashore of the ship and then the scrubbing, scrapping, burning, replacing and resealing of every inch of its submerged timbers. Though this labour was taxing and dangerous for the pirate, the build-up of '[w]eeds and barnacles adhering to the underside of a hull could seriously impede the speed of a vessel by as much as three knots' (Konstam 2003, 5), so it was undoubtedly worth the effort. Every ship at sea for long periods needed to countenance this task on a regular basis, but the various pressures exerted on merchant and naval fleets meant that this was sometimes impossible; indeed, for much of the period under consideration, naval captains were 'forbidden to careen their ships, due to the expense', despite the fact that they knew 'this was essential if they were to catch pirates' (Earle 2004, 186). Writing to his superiors in 1683, one navy captain reported that 'the pirate sailed three feet to his one' (CSPC 1681–85, Item 963). William Dampier mentions careening his piratical vessels at least a dozen of times in his journal, in one passage, explaining the rationale: 'We now concluded to Careen our ships as speedily as we could, that we might be ready to intercept this fleet' (Dampier 1699, 171). The Navy eventually recognised the importance of this process, ensuring as part of their measures in the 1720s to tackle the piracy problem that 'orders were given to careen twice and then three times a year' (Earle 2004, 187).

As well as moving fast, the practice of piracy demanded stealth. Knowledge of the seas – currents, winds, local geography – clearly mattered in this respect, but one of the ways pirates were able to get close involved a significant, if structurally minor, modification to the ship, involving perhaps the most famous symbol of piracy there is. Like privateers and naval ships, pirates were well acquainted with the *ruse de guerre* of flying false flags, a form of trickery used to advance on prey. The court presiding over the trial of the crew of the infamous Bartholomew Roberts heard of his geopolitically savvy use of flags and colour: 'The Colours they fought under (beside the Black Flag) were a red *English* Ensign, a King's Jack, and a Dutch Pendant' (Anon 1723, 5). Twenty years earlier, in the Indian Ocean, pirates were doing something remarkably similar:

> a Pyrate came into Calcutta Road the 23rd November 1696 under English Colours, where were several ship at anchor, coming in call of the Outermost The Pyrate fired a Gun at her, and hoists Danes Colours; firing broadsides and volleys, small shot. (IOR/H/36, ff.277)

Alongside the various national flags present aboard each pirate ship, there was often a black or red flag, famously (though not always) adorned with 'a death's head and bones' (Anon 1718, 24), created and flown as a symbol of the pirates' violent intent and avowed autonomy. These Jolly Roger flags are certainly the most iconic and arguably the most significant acts of material modification undertaken in the crafting of the *pirate* ship. This simple but effective symbolism, denoting 'death, violence, and limited time' (Rediker 2004, 165), was reproduced time and again by pirates throughout the early eighteenth century. The ultimate aim of such flags was, of course, to stop other ships in their tracks, a tactic of immobility. As Adey (2006, 83) reminds us, 'movement may be an action of domination in one circumstance but it may be viewed as an action of resistance in another. Mobility, like power, is a relational thing'. The raising of the Jolly Roger as the ship was in pursuit of its prey both an act of resistance vis-à-vis the state, and an act of domination with respect to the pursued ships' crew who were supposed to see the flag and yield.

The striking of a black flag was synonymous with piratical intention, and was an important indicator of the 'social and cultural differences ... made, negotiated and contested in and through the geographies of the ship' (Lambert, Martins, and Ogborn 2006, 487). It might be thought to signal the kind of social contract articulated in 1721 by one Admiralty judge when he accused the pirates who stood in front of him of the following:

> You did Bind yourselves ... to stand by one another & by your Captain or any other one of them ... to the Last drop of your Blood in your pyrat articles ... Called the Customs of the Blades of Fortune. (AC/16, ff.323)

In the taking of a ship, pirates seized space and seemingly remade it in their image. While the clearing and careening of the ship speak of a clear desire to move quickly, the raising of false flags and the notorious Jolly Roger is more about stealth, about getting close and sending a visual warning to other ships that they should not attempt to engage or flee. These are clear practices of on-going ship alteration bound up with the specificities of pirate life, which serve as important markers of the distinctiveness of the pirate ship within the wider notion of 'the ship', and prompt us to see the ship both materially and politically as a dynamic rather than fixed object.

Making Things Flush: Speed or Sedition?

Early modern seafarers knew all too well that ships were often sites of division and difference:

> By the eighteenth century, the quarterdeck was sacred to the presence of sovereign power in displays of etiquette and privilege. It was the captain's territory – his to walk on alone, his to speak from but not to be spoken to unless he wished it ... The quarterdeck embodied his commission from the King. It was the space of his sovereign's power, and all its trivial gestures and etiquette were its geography. (Dening 1992, 19)

Ships built for navies and traders were designed with specific social geographies in mind, principally targeted at separating officers and sailors in their work, rest and play, and the arrangement of decks was central to this. Such demands on space could only ever be intentions, the reality of socio-spatial relations is that space is contested and negotiated in an on-going, processual way. There have been tentative suggestions that we might consider the ship to be an 'immutable mobile' (e.g. Law 2002; Hetherington and Law 2000), indeed Law and Mol (2001, 612) ask the following question: 'Is there no change in the working relations between the hull, the spars, the sails, the sailors and all the rest? If this is the case then the ship is immutable in the sense intended by Latour. It does not move in relation to a network space'. One of the main arguments being pursued in this paper is that any exploration of the mobilities and spatialities of ships needs to reckon with the fact that many ships are mutable rather than immutable mobiles. As Jarvis (2007, 55) puts it: 'Far from being static entities, ships were organically and often repeatedly altered over the course of their life spans'.

Aboard pirate ships, and other ships in different ways for different reasons, the material form of the ship *did* change as it moved, in response to practical concerns about stowing cargo, natural phenomena such as storms and social pressures such as mutiny. Examples from beyond the pirate ship confirm this. In her work on slave ships, Webster (2008, 7) has noted that a 'vessel engaged on a slaving venture was not, for much of its voyage, a "slave ship" at all'. Throughout its circulation, making its way around the Atlantic triangle, the 'slave ship' required a 'routine series of modifications – mainly carried out as vessels lay off the coast of Africa negotiating for slaves – that transformed merchant ships designed to transport inanimate cargoes into slave ships designed to transport, under duress, cargoes of human beings' (Webster 2005, 250). More generally, Jarvis (2007, 55) insists that:

> Ships were readily modified … and structurally enhanced (rerigged, hold configurations altered, physically expanded vertically and laterally) as needs arose to fit better, new, or different maritime uses … Form and fabric were thus changeable, and descriptions must be situated in time for a given vessel.

One way of demonstrating this mutability is to consider a little acknowledged spatial modification noted in a number of accounts of the pirate ship; namely, the shifting of the decks to make the ship flush. With years of experience in 'legitimate' maritime services behind them, sailors turned pirates undoubtedly understood that ships were intended to be spatially divisive, but that this was subject to change and modification. It, therefore, follows that we might see their resistance to this hierarchy – should there be any – in the way they chose to structure, or re-structure, the pirate ship.

In 1721, the mutinous crew of the *Gambia Castle* set about reworking the now former slaving vessel to suit better their piratical ambitions, and in their first autonomous act, they 'knocked down the cabins, made the ship flush for[e] and aft' before they 'prepared black colours' (Johnson 1726, 307). John Gow's pirate ship was also noted as being 'flush fore & aft' (SP 54/15/3b, ff.10; 11) and at around the same time we find yet that another crew endeavouring to make much the same arrangements aboard their own prize:

> The pirates kept the *Onslow* for their own use … and then fell to making such alterations as might fit her for a sea-rover, pulling down her bulkheads, and making her flush, so that she became, in all respects, as complete a ship for their purpose as any they could have found. (Johnson 1726, 229)

Their labours in reworking the form of the ship gave them the ideal pirate ship. Amongst other things, their ideal entailed a flush ship, an end which they could have achieved, in some vessels at least, if they 'removed the forecastle and lowered the quarterdeck' to ensure that the vessel was 'without a break or step in the weather deck' (Cordingly 1995, 159). This was certainly the case when the pirate John Martel apparently implored the captain of the captured *John and Martha* to 'tell his

Owners that their Ship would answer his Purpose exactly, by taking one Deck down' (Johnson 1726, 64). Again in the *General History of the Pyrates*, we find that upon taking a vessel in the river Gambia, the commanding pirate 'went into her, with his Crew, and cutting away her half Deck, mounted her with 24 Guns' (Johnson 1726, 175).

While it is impossible to tell quite how common this undertaking was – pirates are not noted for leaving substantial paper trails – it is apparent that it was not at all uncommon. In an account of his travels among pirates in the 1680s, Basil Ringrose (Sloane MSS 3820, pg. 71; 84; 171) relates that he and his fellow crew 'tooke downe our roundhouse and coache and all the high carved works of her sterne' and found that the ship sailed 'much better for her alteration'; a development which they later improved upon further by removing one deck. Dampier (1699, 380) also comments on a similar decision made by the crew in which he was part of, to instruct the 'Carpenters to Cut down our Quarter-Deck to make the Ship snug, and the fitter for sailing'. A contemporary of both these men, John Cox, records a remarkably similar process in his own journal: 'Our Ship had 2 decks & large quarter deck ... we cut off her upper deck ... & lowered her quarterdeck [convenient] for a great Cabbin & this work we did all in 10 days time' (GOS/4, 46). The task was not insignificant and clearly required a degree of skill and a deep understanding of maritime architecture. Almost 30 years later, a mutinously minded sailor called Robert Sparkes articulated a very similar thought process while serving aboard the slaving ship *Abingdon*. He whispered to some of his fellow sailors that 'he believed that he could make the Ship go much better than she did' by 'Ripping off the upper Deck', thus making it 'a Ship fit for Business', claiming that 'she would make a good Pirate Ship, for he believed, that she would be stiffer and go better' (Anon 1721, 40).

In a number of the examples cited here, it is clear that speed and increased mobility was a motivating factor in the process of removing the upper deck of the pirate ship. If these pirates are understood as purely economic creatures (cf. Leeson 2009), then this functional 'speed' explanation suffices. However, since such explanations fail to capture the complexity of piracy in every other respect, then it follows that they would also fall short of fully explaining why pirates would alter the spatial arrangement of their ships.

Another way of interpreting this practice is hinted at by Baer (2005, 208) when he writes that the removal of the 'upper work on a pirate ship' was 'primarily to improve its agility but also to eliminate class differences among its crew'. Beyond this paper, Baer's suggestion remains unchallenged and unsubstantiated in the literature, despite the fact that it chimes so clearly with one of Rediker's (2004, 65) central claims about piracy; that aboard the pirate ship we can see the 'determined reorganisation of space and privilege [which] was crucial to the remaking of maritime social relations'. Developing our understanding of the pirate ship requires critical attention to ways in which that space was created – the ways that the ship metamorphosed into the pirate ship – and the ways in which it was then inhabited. By focusing on these hitherto ignored fluidities and processes, we are able to gain new perspectives on 'space itself and how it is produced (and reproduces itself) within the dynamics of spatial assemblages' (Steinberg 2013, 163). What is more, focusing on these mobile materialities allows us to really grapple with the social and political geographies of pirate ships, unsettling those narratives, such as that of the radical pirate (Rediker 2004) that have failed to take them seriously (cf. Merriman et al. 2012).

Looking for the Complex Geographies and (Im)mobilities of the Pirate Ship

Jensen (2009, xv) rightly insists that 'mobile practices are more than physical practices, as they also are signifying practices', so as well as foregrounding the materialities of pirate ships – highlighting their mutability – this paper brings this transmutative space into conversation with the complex social and political geographies to be found within. The pirate ship of the early eighteenth century was, Rediker (2004, 61) argues, the scene for a 'new social order … conceived and deliberately constructed by the pirates themselves'. There is plenty of evidence to support this. Rodgers (1718, xv), a colonial governor, noted that 'there was no distinction between the Captain and Crew: for *the Officers having no Commission but what the Majority gave them*, were chang'd at every Caprice'. Similarly, the merchant William Betagh (1728, 148) testified that pirates 'had no regular command among them'. After an encounter with pirates aboard their ship, another merchant noted that: 'There appeared Very Little order amongst the crew and that every one of them dealt about Wine and fruits to persons who came aboard at their pleasure' (SP/15/3b, ff.11). What these men all mistook for disorder was, Rediker (2004, 61) argues, merely a 'different social order – different from the ordering of merchant, naval and privateering vessels'.

Such a profound difference ought to be manifest in spatialities and mobilities of the ship. Can we read the examples of deck alteration cited above as evidence of pirates' egalitarian will manifest in the re-organisation of the ship? Among the things that captured the attention of Snelgrave during his time among the pirates was their apparent lack of acquiescence to the authority of their captain. This extended to sleeping arrangements, which involved the entire crew having to 'lay rough ... that is, on the deck; the Captain himself not being allowed a Bed' (Snelgrave 1734, 217). Where someone like Snelgrave (1734, 216) would expect a captain to sleep, he found an altogether different scene: 'There was not in the Cabbin either Chair, or anything else to sit upon ... [only] a Carpet was spread on the Deck, upon which we sat down cross-legg'd'. Another observer at close quarters with pirates noted the same arrangement, writing that 'the Captain cannot keep his own Cabbin to himself, for ther Bulk-Heads are all down, and every Man stand to his Quarters, where they lie and mess, and they take the liberty of ranging all over the ship' (Downing 1737, 108). These examples seem to suggest that the space of the pirate ship, and the lack of barriers to movement between its different areas, did reflect the supposedly egalitarian values of the pirates, in some cases.

Different pirate ships were different spaces. Nathaniel North is said to have received his captaincy of a pirate ship like this:

> The Ceremony of this Installation is, the crew having made choice of him to Command ... desire he will take upon him the Command, as he is the most capable among them. That he will take Possession of the great Cabin; and on accepting the Office, he is led into the Cabin in State, and placed at a Table, where only one Chair is set at the upper End, and one at the lower End of the Table for the Company's Quarter-Master. (Johnson 1726, 524)

A similar tone is apparent in the bloody rise of William Fly to the position of captain of the *Elizabeth* in 1726: 'All Obstacles being removed, *Mitchell* saluted *Fly* Captain and, with the rest of the Crew, who had been in the Conspiracy, with some

Ceremony, gave him Possession of the great Cabin' (Johnson 1726, 608). Of Captain Davis, Johnson (1726, 193) writes that he 'held Certain Privileges, which common Pyrates were debarr'd from, as walking the Quarter-Deck, using the great Cabin, going ashore at Pleasure'. When Snelgrave (1734, 240) was aboard Davis' ship, he recalled that the pirate captain 'spoke to all his people on the Quarter-deck', hinting at how he occupied it *as* captain. This situation is further demonstrated in another of Snelgrave's recollections; namely, that on a particular occasion when one of the pirate crew approached him aggressively, in protection of his captive, 'Davis caned him [the approaching pirate] *off the Quarter-deck*' (Snelgrave 1734, 274). More striking still is an account of Gow's immediate spatial organisation of the ship he had seized:

> THE first Order they Issued, was to let all the rest of the Men know, That if they continued Quiet, and offer's not to Meddle with any of their Affairs, they should receive no Hurt: But strictly forbid any Man among them to set a Foot Abaft the Main-mast, except they were call'd to the Helm, upon Pain of being immediately Cut in Pieces; keeping for that Purpose, one Man at the Steerage-door, and one upon the Quarter-deck, with drawn Cutlashes in their Hands. (Anon 1725, 10–11)

Besides offering a window into the often brutal world of the pirate ship, this somewhat hyperbolic observation highlights a key concern of this paper, namely that an analysis of the materialities and mobilities of the pirate ship seems to reveal that the crew of the pirate ship and 'the humanity of the ship' were not one and the same; the former was the dominant part of the latter.

The geographies of the pirate ship were complex and did not straightforwardly flatten out spatialised hierarchies. It was noted earlier that Gow's ship was one which was flush, but here we have a divided ship. Perhaps the ship changed form over time, perhaps they changed ships, or it could be that one of the recorded observations is unreliable. It remains, however, that following Massey (2005, 99), we can see the space of the ship better as 'the sphere of heterogeneity ... of relations, negotiations, practices of engagement, power in all its forms ... of coevalness, of radical contemporaneity'. The pirate ship was a space in flux, socially and materially on the move, always becoming. Taking this approach addresses Buck-Morss' (2009, 79) call to work:

> ...through the historical specificities of particular experiences, approaching the universal not by subsuming facts within overarching systems or homogenizing premises, but by attending to the edges of systems, the limits of premises, the boundaries of our historical imagination in order to trespass, trouble, and tear these boundaries down.

Approaching the pirate ship through the lens of spatiality and mobility embraces Buck-Morss' argument by foregrounding questions about the mutable socio-spatialities of the pirate ship. Such questions take us to the edges of both conservative (Earle 2004) and radical (Rediker 2004) histories of piracy, to the limits of premises of avaricious brutality and anti-authoritarianism. Following Massey, Buck-Morss, and others towards thinking about the pirate ship as a space of heterogeneous relations, as a space comprising more than simply 'the crew', opens up the possibility for more nuanced interpretations of piracy and pirate ships. Unfixing the pirate ship, stripping the metaphor bare, reveals a more complex and challenging set of agents, motives, forces and relations.

As Featherstone (2008, 109) has argued, '[f]oregrounding the forms of political activity constituted through these shipboard spaces offers possibilities for asserting and allowing dynamic trajectories of subaltern political identities'. Pirates are, of course, one subaltern group, but the pirate ship contained other subalterns who are somewhat lost in most accounts of the pirate ship. With space limited, I can only focus on one of these groups and their place aboard the pirate ship to explore the implications of working with changing spatialities and mobilities.

Despite it being the apparent 'Custome among the Pyrates to force no Prisoners', in a number of trials, and in the accounts of many observers, the (mis)use of prisoners among pirate crews is discussed (Anon 1717a). When considering the treatment of these people, who lived for a period aboard the pirate ship, but were not part of its crew, it is important to consider how space and (im)mobility shaped their captivity. For example, in 1696, one merchant reported his captivity among pirates to the Board of Trade, informing them that: 'They used us extremely hard, beat us, pinched us of victuals, *shut us down in the night to take our lodgings in the water-casks*' (CSPC 1696–97, Item 1203: emphasis added). Another sailor testified to the Admiralty in 1702 that the pirates who seized his ship 'did also seize the men that belonged to ye ship … and put them under a guard' (HCA 1/53, f. 133). Without much empathy, one Admiralty official noted during a trial in 1720, that there was a tendency among pirates to 'force' men:

> That they themselves were taken at Sea by the **Pirate-Crew** while sailing … and were compelled to go along with the **Pirates** by Threats, and the Apprehension they had of the Treatment they might meet with from so **barbarous a Set**, in case of Refusal … That they at different Times endeavoured severally to make their Escape by running away, and were brought back again, and some of them whipt for so endeavouring to make their Escape, and others kept for some Time in Irons after. (AC 9/681, F.3)

In the aforementioned case of Gow, an account of his piracies holds that of the 24 crew members aboard the ship he seized, only eight were active in the mutiny, meaning that sixteen were unwilling. Four more joined with the pirates shortly afterwards, leaving twelve aboard who did not want to be part of the pirate crew, but were nevertheless taken (Anon 1725, 11). From 1696 onwards the Admiralty Court took the line that to be aboard a pirate ship was to be a pirate, unless the accused could prove otherwise. Regular acquittals in pirate trials show, however, that there was clearly some acknowledgement that pirate ships contained a 'crew' *and* 'others'.

Of particular interest to pirate crews were those possessing knowledge and skill in carpentry and medicine, both potentially life-saving talents afloat. Thomas Davis was a carpenter and, when his ship was captured in 1717 by pirates, he was coveted by the crew of the pirate ship *Whydah*, so much so, that they would rather have him dead than not in their company. When the *Whydah*'s captain 'asked his Comp'y if they were willing to lett Davis the Carpenter go', they 'Expressed themselves in a Violent manner saying no, Dam him, they would first shoot him or Whip him to Death at the Mast' (Anon 1717a). The ship aboard which Davis might have been whipped to death was formerly a slaving vessel, and so corporeal punishment of crew and cargo at the mast would have been a familiar event, especially for those among the pirate crew who had been slave-ship mariners themselves. In the transition, things changed dramatically aboard the pirate ship, but not all of the oppressive structures and practices were removed. The mast was itself a site of symbolic importance aboard the ship, a sort of public place where the collective will for punishment could be exercised, where individuals felt the full force of not being part of that collective. In the same run of trials, John Brett, related to the court that one of the pirate crew, a certain John Brown had 'told a prisoner then on board that he would hide him in the hold, and hinder him from Complaining against him, or telling his Story' (Anon 1717b). Acknowledging the relational nature of space, mobility and power *within the ship* through attention to the mutable materialities of the ship brings these souls into our picture of piracy and gives them agency. Those marginalised by the re-ordered but still ordered pirate ship, those whipped at the mast or bundled into the hold to prevent them from seeking *their own* path, their own idea of liberty, are revealed when we see the ship as process rather than outcome.

An ever-more intriguing picture of the pirate ship emerges from such a position. One prisoner of pirates told a court that he was abused by some and not by others, suggesting that social relations (and spatial order) aboard were never set but were a matter of negotiation and contestation. Pirate ships were, like all other spaces, sites of juxtaposition wherein 'contemporaneous heterogeneities' (Massey 2005, 5) existed in tension. Seeing the ship as a mutable mobile brings to light an ever-evolving distinction between the crew, who were signed up to the pirates' articles and enfranchised, and others, who, for whatever reason, were not part of the crew. It is the latter group who have so often been lost from analyses of piracy, an inconvenient element left at the fringes of the narrative. Spatialities and mobilities unsettle these narratives to capture important juxtapositions, allowing us to more sensitively incorporate these 'non-pirate' inhabitants of the pirate ship into the picture of piracy, perhaps in a fashion which escapes the rather grand narratives of historians bent on telling stories of naval dominance and piratical resistance. The pirate ship was, in fact, a site wherein a 'plurality of trajectories' (Massey 2005, 5) took shape and flight, moved and were moved, a site of contestation wrought in shape-shifting timbers and personal politics, a space where power was shared differently but always unevenly.

Conclusions

Piracy begins with the ship, with the deliberate and forceful appropriation of space afloat. Seditious thoughts, rebellious whispers and violent actions became piracy when they culminated in the taking or attempted taking of a ship. 'Ships are first and foremost physical objects and artefacts belonging to particular times, places, and cultures' (Jarvis 2007, 52), and while the act of piracy is deeply infused with symbolic and political meaning (Rediker, 2004), it is, first and foremost, a material manoeuvre. This paper has sought to foreground some of the ways that the pirate ship existed as a real, lived and dynamic space, one crafted by pirates in their own image with their own ends in mind. The ship functioned as a technology of mobility and speed, as a locale for piratical politics and as a space of multiple contestations, and revealing their spatial practices in modifying this space sheds much needed light on their intriguing way of life. The practices of pirates in converting ships to pirate ships have been discussed as a means of demonstrating the inherently unstable nature of ship space, its mutability. Through this discussion, it has been shown that the pirate ship is a far more complex place, both socially and materially, than is commonly thought. This complexity challenges dominant views of the pirate ship as an inherently radical or anarchic site by allowing for the processual nature of space and mobility, suggesting that social relations were contested and negotiated in an ongoing way, rather than one which is fixed in time and space. Current understandings of the pirate ship – and perhaps many other kinds of ships besides – hold only so far as we can hold space still.

Using records of pirate's shipboard activities, this paper not only addresses issues in the historiography of piracy, it also begins to address some of the gaps in new mobilities paradigm. As well the need for more consideration of 'watery mobilities', there is also a pressing need for 'a strong sense of historical consciousness' in mobilities studies (Cresswell 2011, 555). As Ashmore (2013, 1) has also contended, the long-distance or long-term journeying of seafarers, whether as passengers, pilots or pirates, is 'a form of mobility that has received limited attention in the mobilities literature'. Furthermore, developing our understanding of the historical geographies and mobilities of the sea, 'necessarily involves thinking also about ships and the spaces on board ships' (Ryan 2006, 580). This exploration of the mobile and mobilising form of the pirate ship develops these emerging debates, adding a maritime angle and historical perspective to the literature which contributes to our understanding of mobilities as much as it does our understanding of pirates, ships and the seafaring world. As Steinberg (2013, 160) has argued, 'objects come into being as the move (or unfold) through space and time', so this focus on mobilities furthers our appreciation of the world of the pirate through an attention to the materialities of their lives afloat. Change and transformation, as Hannam, Sheller, and Urry (2006, 14) have suggested must be central to our interrogation of materialities, mobilities and space: 'there is now a growing interest in the ways in which material "stuff" makes up places, and such stuff is always in motion, being assembled and reassembled in changing configurations'. This paper has embraced this challenge and adds to our understanding of just how piracy worked as a mobile, material and spatial practice afloat, showing some of the ways that mobilities theory might be used to further interrogate the mobile worlds of the ship and the seafarer.

Acknowledgements

The arguments presented in this paper derive from doctoral work funded by the ESRC, and have benefited considerably from comments and criticisms offered by Chris Philo and Dave Featherstone, as well as participants at the 'Being in Transit' workshop in Heidelberg, and the RGS-IBG 'Geographies of Ships' session, and *Mobilities*' editors and reviewers.

Notes

1. Though the term 'metamorphosis' denotes a very specific physical process in the biological sciences, the word also conveys rather neatly the idea of radical material or physical change. The etymology of the word is revealing in this regard, coming from the Greek meaning 'to transform, change shape'.
2. For a sustained critical discussion of metaphor in relation to notions of space and mobility, see Urry (2000).

References

Primary Sources

AC 16/1. *Criminal Records, 5 March 1705-24 September 1734*. Edinburgh: National Archives of Scotland.

AC 9/681. *Admiralty Court Processes in Foro, 1703–1830*. Edinburgh: National Archives of Scotland.

Anon. 1717a. *Trial of Thomas Davis, October 28th 1717 (Suffolk Court Files, Fragment 99)*. Reprinted in Jameson, J. F., ed. 1970. *Piracy and Privateering in the Colonial Period: Illustrative Documents*. New York: A.M. Kelley.

Anon. 1717b. *Trial of Simon van Vorst and Others, October, 1717 (Suffolk Court Files, no 10923; a Fragment)*. Reprinted in Jameson, J. F., ed. 1970. *Piracy and Privateering in the Colonial Period: Illustrative Documents*. New York: A.M. Kelley.

Anon. 1718. *The Trials of Eight Persons Indicted for Piracy & c. Of Whom Two were Acquitted, and the Rest found Guilty. Boston, John Edwards, 1718. [Gottingen Library, Shelfmark MC 83-1021:2003]*. Reprinted in Baer, J. 2007. *British Piracy in the Golden Age: History and Interpretation, 1660–1730*. London: Pickerington & Chatto.

Anon. 1721. *The Tryals of Captain John Rackam, and Other Pirates ... Who were All Condemn'd for Piracy, at the Town of St. Jago de la Vega, in the Island of Jamaica, on Wednesday and Thursday the Sixteenth and Seventeenth Days of November 1720. As also, the Tryals of Mary Read and Anne Bonny, Alias Bonn. Jamaica, 1721. [National Archives, Shelfmark CO 137/14/9-30]*. Reprinted in Baer, J. 2007. *British Piracy in the Golden Age: History and Interpretation, 1660–1730*. London: Pickerington & Chatto.

Anon. 1723. *A Full and Exact Account, of the Tryal of all the Pyrates, Lately Taken by Captain Ogle, on Board the Swallow Man of War, on the Coast of Guinea. London: Printed, and Sold by J. Roberts, 1723*. Edinburgh: National Library of Scotland.

Anon. 1725. *Anon, An Account of the conduct and proceedings of the late John Gow alias Smith, captain of the late pirates... London: Printed and sold by John Applebee, in Black-Fryers, [1725]*. Eighteenth Century Collections Online, ESTC # T056855.

Betagh, W. 1728. *A Voyage Round the World ... London : Printed for T. Combes at the Bible and Dove in Pater-Noster Row, J. Lacy at the Ship Near Temple Bar, and J. Clarke at the Bible under the Royal Exchange, MDCCXXVIII. [1728]*. Eighteenth Century Collections Online, ESTC # T181354.

Blum, H. 2010. "The Prospect of Oceanic Studies." *Publications of the Modern Languages Association of America* 125 (3): 670–677.

CSPC. 1681–85. *Calendar of State Papers, Colonial Series, America and the West Indies, 1681–1685* [Edited by J. W. Fortescue, Printed for Her Majesty's Stationary Office by Eyre and Spotiswood, London, 1898.

CSPC. 1696–97. *Calendar of State Papers, Colonial Series, America and the West Indies, 1696–1697* [Edited by J. W. Fortescue], Printed for His Majesty's Stationary Office by Mackie and Co, London, 1904.

Dampier, W. 1699. *A New Voyage Round the World..., London, 1699*. Glasgow: University of Glasgow Special Collections.

Downing, C. 1737. *A Compendious History of the Indian Wars ... with an Account of the Life and Actions of John Plantain, a Notorious Pyrate at Madagascar, London, Printed for T. Cooper, at the Globe in Pater-noster Row, 1737*. Edinburgh: National Library of Scotland.

GOS/4. *Logbook Kept by John Cox 'His Travaills Over the Land into the South Seas from Thence round the South Parte of America to Barbadoes and Antigua', 1680–82*. London: National Maritime Museum.

HCA 1/53. *Examinations of Pirates and Other Criminals, 1694–1710*. London: National Archives.

IOR: H/36. *Indian Office Records, Miscellaneous Papers, 1658–1699*. London: British Library.

Johnson, C. 1726. *"History of the Pyrates...", 1726, London*. Reprinted in Schonhorn, M., ed. 1999. *A General History of the Pyrates*. London: Dover.

Rodgers, W. 1718. *A cruising Voyage Round the World ... London: Printed for Andrew Bell at the Cross-keys and Bible in Cornhil, and Bernard Lintot at the Cross-Keys between the Temple-Gates, Fleetstreet, M.DCC.XVIII. [1718]*. Eighteenth Century Collections Online, ESTC # T131767.

Sloane MSS. 3820. *Maps and Plans: Maps Belonging to B. Ringrose's Journal of Capt. B. Sharp's Voyage to the South Seas: 1680–1682. South Seas: Journal of voyage to, by B. Ringrose: 1680–1682. Bartholomew Sharp, Captain: Journal of His Voyage to the South Seas, by B. R.* London: British Library.

Snelgrave, W. 1734. *A New Account of Some Parts of Guinea, and the Slave-trade ... A Relation of the Author's being Taken by Pirates, and the Many Dangers he Underwent. By Captain William Snelgrave. London: printed for James, John, and Paul Knapton, at the Crown in Ludgate Street, MDCCXXXIV. [1734].* Eighteenth Century Collections Online, ESTC # T136167.

SP 54/15. *Secretaries of State: State Papers Scotland Series II.* London: National Archives.

Secondary Sources

Adams, J. 2001. "Ships and Boats as Archaeological Source Material." *World Archaeology* 32 (3): 292–310.

Adey, P. 2006. "If Mobility is Everything Then It is Nothing: Towards a Relational Politics of (Im) mobilities." *Mobilities* 1 (1): 75–94.

Anim-Addo, A. 2011. "'A Wretched and Slave-like Mode of Labor': Slavery, Emancipation, and the Royal Mail Steam Packet Company's Coaling Stations." *Historical Geography* 39: 65–84.

Ashmore, P. 2013. "Slowing Down Mobilities: Passengering on an Inter-war Ocean Liner." *Mobilities* 8 (4): 595–611.

Baer, J. 2005. *Pirates of the British Isles.* Gloucestershire: Tempus.

Benton, L. 2010. *A Search for Sovereignty: Law and Geography in European Empires, 1400–1900.* Cambridge: Cambridge University Press.

Bialuschewski, A. 2008. "Black People Under the Black Flag: Piracy and the Slave Trade off the West Coast of Africa, 1718–1723." *Slavery and Abolition* 29 (4): 461–475.

Boshier, R. 2009. "Wet and BOISTEROUS: The Lumpy 'Romance' of Commuting by Boat." In *The Cultures of Alternative Mobilities: Routes Less Travelled*, edited by P. Vannini, 195–210. Farnham: Ashgate.

Buck-Morss, S. 2009. *Hegel, Haiti, and Universal History.* Pittsburgh, PA: University of Pittsburgh Press.

Casarino, C. 2002. *Modernity at Sea: Melville, Marx, and Conrad in Crisis.* London: University of Minnesota Press.

Chamber, I. 2010. "Maritime Criticism and the Lessons from the Sea." *Insights* 3 (9): 1–11.

Cordingly, D. 1995. *Under the Black Flag: The Romance and the Reality of Life among the Pirates.* London: Harvest.

Cresswell, T. 2010. "Towards a Politics of Mobility." *Environment and Planning D: Society and Space* 28 (1): 17–31.

Cresswell, T. 2011. "Mobilities I: Catching Up." *Progress in Human Geography* 35 (4): 550–558.

della Dora, V. 2010. "Making Mobile Knowledges: The Educational Cruises of the Revue Générale des Sciences Pures et Appliquées, 1897–1914." *Isis* 101 (3): 467–500.

Dening, G. 1992. *Mr Bligh's Bad Language: Passion, Power and Theatre on the Bounty.* Cambridge: Cambridge University Press.

Earle, P. 2004. *The Pirate Wars.* London: Methuen.

Featherstone, D. 2005. "Atlantic Networks, Antagonisms and the Formation of Subaltern Political Identities." *Social and Cultural Geography* 6 (3): 387–404.

Featherstone, D. 2008. *Resistance, Space and Political Identities: The Making of Counter-global Networks.* Oxford: Wiley-Blackwell.

Foucault, M. 1986. "Of Other Spaces." *Diacritics* 16 (1): 22–27.

Gilroy, P. 1993. *The Black Atlantic: Modernity and Double Consciousness.* London: Verso Press.

Gosse, P. 1924. *The Pirates' Who's Who: Giving Particulars of the Lives and Deaths of the Pirates and Buccaneers.* London: Dulau & Co.

Hannam, K., M. Sheller, and J. Urry. 2006. "Editorial: Mobilities, Immobilities and Moorings." *Mobilities* 1 (1): 1–22.

Hasty, W. 2011. "Piracy and the Production of Knowledge in the Travels of William Dampier, c.1679–1688." *Journal of Historical Geography* 37 (1): 40–54.

Hasty, W., & Peters, K. 2012. "Geographies of the Ship and the Ship in Geography." *Geography Compass* 6 (11): 660–676.

Hill, C. 1972. *The World Turned Upside Down: Radical Ideas During the English Revolution*. London: Penguin.
Jarvis, M. 2007. "On the Material Culture of Ships in the Age of Sail." In *Pirates, Jack Tar, and Memory: New Directions in American History*, edited by P. A. Gilje and W. Pencak, 51–72. Connecticut: Mystic Seaport.
Jensen, O. B. 2009. "Foreword Mobilities as Culture." In *The Cultures of Alternative Mobilities: Routes Less Travelled*, edited by P. Vannini, xv–xix. Farnham: Ashgate.
Konstam, A. 2003. *The Pirate Ship 1660–1730*. Oxford: Osprey.
Lambert, D., L. Martins, and M. Ogborn. 2006. "Currents, Visions and Voyages: Historical Geographies of the Sea." *Journal of Historical Geography* 32 (3): 479–493.
Law, J. 2002. "Objects and Spaces." *Theory, Culture and Society* 19 (5–6): 91–105.
Law, J., and K. Hetherington. 2000. "Materialities, Spatialities and Globalities." In *Knowledge, Space, Economy*, edited by J. R. Bryson, P. W. Daniels, N. Henry, and J. Pollard, 34–49. London: Routledge.
Law, J., and A. Mol. 2001. "Situating Technoscience: An Inquiry into Spatialities." *Environment and Planning D: Society and Space* 19 (5): 609–621.
Leeson, P. T. 2009. *The Invisible Hook: The Hidden Economics of Pirates*. Oxford: Princeton University Press.
Linebaugh, P., and M. Rediker. 2000. *The Many-headed Hydra: Sailors, Slaves, Commoners, and the Hidden History of the Revolutionary Atlantic*. Boston, MA: Beacon Press.
Massey, D. 2005. *For Space*. London: Sage.
Merriman, P., M. Jones, G. Olsson, E. Sheppard, N. Thrift, and Y.-F. Tuan. 2012. "Space and Spatiality in Theory." *Dialogues in Human Geography* 2 (1): 3–22.
Pearson, M. N. 2010. "The Idea of the Ocean." In *Eyes Across the Water: Navigating the Indian Ocean*, edited by P. Gupta, I. Hofmeyr, and M. Pearson, 7–14. Pretoria: Unisa Press.
Peters, K. 2012. "Manipulating Material Hydro-worlds: Rethinking Human and More-than-human Relationality through Offshore Radio Piracy." *Environment and Planning A* 44 (5): 1241–1254.
Urry, J. 2000. *Sociology Beyond Society: Mobilities for the Twenty-first Century*. London: Routledge.
Rediker, M. 1987. *Between the Devil and the Deep Blue Sea*. Cambridge: Cambridge University Press.
Rediker, M. 2004. *Villains of All Nations: Atlantic Pirates in the Golden Age*. London: Verso.
Ryan, J. R. 2006. "'Our Home on the Ocean': Lady Brassey and the Voyages of the Sunbeam, 1874–1887." *Journal of Historical Geography* 32 (3): 579–604.
Serres, M. 1995. *The Natural Contract*. Ann Arbour: University of Michigan Press.
Sheller, M., and J. Urry. 2006. "The New Mobilities Paradigm." *Environment and Planning A* 38: 207–226.
Smith, N., and C. Katz. 1993. "Grounding Metaphor: Towards a Spatialised Politics." In *Place and the Politics of Identity*, edited by S. Pile, 66–81. London: Routledge.
Steinberg, P. E. 2013. "Of Other Seas: Metaphors and Materialities in Maritime Regions." *Atlantic Studies* 10 (2): 156–169.
Strohmayer, U. 2011. "Bridges: Different Conditions of Mobile Possibilities." In *Geographies of Mobilities: Practices, Spaces, Subjects*, edited by T. Cresswell and P. Merriman, 119–135. Farnham: Ashgate.
Vannini, P. 2011. "Constellations of Ferry (Im)mobility: Islandness as the Performance and Politics of Insulation and Isolation." *Cultural Geographies* 18 (2): 249–271.
Virilio, P. 1977. *Speed and Politics: An Essay on Dromology*. Los Angeles, CA: Semiotext(e).
Webster, J. 2005. "Looking for the Material Culture of the Middle Passage." *Journal for Maritime Research* 7 (1): 245–258.
Webster, J. 2008. "Slave Ships and Maritime Archaeology: An Overview." *International Journal of Historical Archaeology* 12 (1): 6–19.

'The Great Event of the Fortnight': Steamship Rhythms and Colonial Communication

ANYAA ANIM-ADDO

School of History, University of Leeds, Leeds, England

ABSTRACT *This paper engages with Tim Cresswell's 'contellations of mobility' in order to contribute some understanding of historical maritime rhythms. The empirical focus is upon a steamship mail service in the post-emancipation Caribbean. In examining this communications network, it is stressed that while those managing the network valorised predictable efficiency, 'friction' was prized by mercantile groups at the steamers' ports of call. Thus, the different aspects of mobility signified differently across the network, and this historical case study reinforces the resonance of slowness and stoppage time. The synchronisation of steamship arrivals with sociocultural norms in the Caribbean colonies also necessitated the adaptation of mail service rhythms. Through a focus on shipping operations, this paper proposes to temper our understanding of the role of steamship technology in empire. The influence of colonies on the metropole encompassed an alteration of the rhythms of imperial circulation, and it is within the maritime arena that these realities came into sharp focus.*

The Royal Mail Steam Packet Company's (RMSPC's) *Dee* departed from Valparaiso in late May 1848 and stopped at Panama before proceeding to Jamaica, Cuba, and several of the Windward and Leeward Islands. The vessel subsequently called at St Thomas in the Danish West Indies on 15 July (*The Standard*, 5 August 1848). The *Dee* arrived in Southampton almost three weeks later on 4 August 1848, bringing with it the usual news despatches. These were published in newspapers in England the following day.

On the occasion of this particular journey, the RMS *Dee* brought to British shores a combination of news and lurid rumours. *The Examiner* of London reported:

> On the 10th July a slave insurrection took place in St Croix, one of the Danish Antilles. They demanded their freedom, which was granted, deposed the

> governor, Von Scholten, rescued the prisoners from prison, and set fire to and destroyed an immense deal of property all over the island. Part of the town was fired. (*The Examiner*, 5 August 1848)

The *Hampshire Advertiser* alternatively recounted:

> The whole of the property of the Danish Island of St. Croix has been destroyed, owing to an insurrection of the Negroes, who deposed the Governor and demanded and obtained their emancipation. About 5000 men were at one time in arms. The insurgents committed frightful excesses; to infuriate themselves they mixed hogs' blood with rum and drank it to excess. (*Hampshire Advertiser & Salisbury Guardian*, 5 August 1848)

In fact, the events reported in the newspapers had been gathering force since the beginning of July 1848, when the enslaved population in St Croix had begun to revolt and make demands for their freedom (Hall 1992, 208–209). As the August newspaper accounts illustrate, the arrival of the mail steamer was an integral part of imperial communications, since these vessels brought news – in this case dramatic news that struck a particular chord in the long wake of the Haitian Revolution (Geggus 1985, 113) – from Caribbean colonies to European spaces.

The arrival and departure of steamers between colonies proved particularly important to the circulation of information. When news of the insurrection reached St Thomas on 6 July, military support was dispatched from that island. From St Thomas, the Royal Mail steamer *Eagle* proceeded to Puerto Rico, where the Captain General dispatched infantry and artillery within five hours of receiving the intelligence (*The Standard*, 5 August 1848). However, contrary winds prevented these troops from reaching St Croix until the uprising had largely subsided and the troops only arrived on the island on Saturday 8 July (*The Standard*, 5 August 1848). Thus, during the first two weeks of the month, revolutionary impulses swept through St Croix, with enslaved individuals gathering, marching, rioting and claiming freedom. Yet the island's incorporation into the steamship network brought a counter-revolutionary impulse. The RMSPC's steamships, in this case, carried news of revolt to Puerto Rico and in this way facilitated a military mobilisation that conflicted with enslaved people's claims to freedom. The case of St Croix starkly illustrates the significance of the steamship timetable, and the rhythms of arrivals and departures within the Caribbean region. These arrivals and departures – the concern of this article – mattered to island lives, and notably to the mercantile community. Firstly, I discuss the significance of historical oceanic rhythms and suggest that the ocean, as a different kind of material space (Peters 2012) produced a set of rhythms which strained against those of the land. This had particular kinds of consequences for projects of empire, which became frustrated and were necessarily altered within watery spaces. Secondly, in 'Smoothing the steamship timetable', my focus is upon the establishment of these steamship rhythms. Subsequently in 'Steamship pauses' and 'The great event of the fortnight', I turn to the intersection between rhythms at sea and those on shore. As these sections will indicate, steamship rhythms prompted island responses and necessary negotiations at mundane as much as at exceptional moments in the nineteenth century Caribbean. Phillip Vannini has examined the maritime rhythms of 'everyday life' through a contemporary focus on ferry mobilities

(Vannini 2012). Within an historical context, the maritime rhythms of imperial projects betray continual negotiation between metropolitan and colonial spaces.

Time-space compression, much discussed in the context of the twenty-first century, has been historically applied to consideration of nineteenth-century technologies (Cresswell 2006, 4). Thomas Eriksen, for example, highlights that steamers and the telegraph were 'innovations [that] changed the perception of space and distance' (Eriksen 2007, 1–2). Such effects bore particular significance for projects of empire; writing of steamboats, Daniel Headrick stresses that 'Few inventions of the nineteenth century were as important in the history of imperialism' (Headrick 1981, 7). The age of steam brought hopes of faster and predictable timetabled journeys as steamers were welcomed partly for their promise of passages of even duration. My focus on the steamship rhythms of one particular company (the RMSPC), and a shift of perspective to its Caribbean ports of call serves as a reminder that past rhythms of communication were largely negotiated. If the 'different temporal experiences of urban life need to be insisted upon and seen at the heart of accounts of modernity', this surely holds for colonial as much as other spaces (Highmore 2002, 173). Through an examination of how steamers became one of 'multiple quotidian rhythms' folded into nineteenth-century life at colonial ports, the articulation of colonial relations and sociocultural norms through the integration of this new communications network becomes apparent (Edensor 2010, 2).

Consideration of steamship rhythms invites comparison with the contemporaneous rhythms of trains, which seemed, in the nineteenth century, to promise the '[a]nnihilation of space and time' (Schivelbusch 1977, 41). Like train carriages, steamship decks and saloons 'blurred boundaries' between public and private space (Bissell 2009, 55). While for elite passengers making the transatlantic crossing, such journeys could be unusual and noted undertakings, the arrival of post and passengers by steamer at island ports of call was 'intimately woven into everyday routine' in colonial port towns (Bissell 2009, 42). Steamships formed part of Carbbean 'island time' in the nineteenth century (Vannini 2012, 101). Crucially, maritime rhythms folded together those of metropolitan and colonial spaces, and thus exploration of historical rhythms of the maritime world allows for a consideration of the workings of empire. This paper's historical focus on 'island time' – both in Britain and the Caribbean allows for analysis of the working realities of imperial projects such as transport services. As Vannini (2012, 101) stresses, '[d]ifferent places move at different paces', and the infrastructural challenges that this posed led to revision of imperial projects.

In highlighting the rhythms of steam, I contend firstly that steamship rhythms were negotiated across spaces; and secondly that different aspects of steamship mobility were valorised differently across the network. I stress that despite the rhetoric of the RMSPC's managers situated in Southampton and London, at ports of call across the network, stoppage time and steamers' pauses were highly valued alongside the potential for speedy journeys. I suggest that a focus on the rhythms of steamship travel helps to nuance our understanding of speed and communication in colonial contexts, and highlights that this was more complex an historical process than a linear trajectory towards annihilating distance with speed. Although mindful of the fact that attention to historical rhythms 'means being limited to observing observations', in examining this maritime activity, my analysis of the RMSPC's service contributes to a necessary 'awareness of the mobilities of the past' (Highmore 2002, 176; Cresswell 2011, 168)

The RMSPC had its origins in imperial concerns, as the transition of the British West Indies from slavery to freedom prompted anti-abolitionist James MacQueen to argue for improved communication between Britain and the Caribbean colonies (MacQueen 1838; Lambert 2008). MacQueen had lived in Grenada managing a sugar estate during the late eighteenth century (Lambert 2008). In the wake of abolition in 1833, MacQueen's concerns for the British West Indies were in alignment with those of planters, who faced the prospect of trying to maintain levels of sugar production with a labour force that they feared would dwindle (Hall 1971, 23). Within the context of anxious debate over the future of the British West Indian colonies, MacQueen secured a British Government mail contract. The mail contract's initial value of £240,000 per year rose to £270,000 when the Company branched into South American operations. Holding high hopes for the transformative power of this shipping network, MacQueen argued that 'The West Indies everywhere want a little European energy and regularity infused into them, – and this is one efficient, perhaps the simplest and most efficient way to do it' (MacQueen 1838, 56). His reference to 'energy and regularity' invites some consideration of the RMSPC's rhythms and their reception in the Caribbean. His words, after all, seem to chime with a view of modernity as 'an insistent and ferocious rhythm' (Highmore 2002, 171). MacQueen hoped, through the steamship service, to contribute to the 'regular, normative rhythms' of the Caribbean (Edensor 2010, 4).

Without being explicitly named as such, mobilities have been a long-standing theme in post-emancipation Caribbean historiography, particularly through scholars' focus on the so-called 'flight' from the estates, or the withdrawal of enslaved workers from plantation spaces after abolition (Hall 1993). For the formerly enslaved, rights over their own movements were an important meaning of freedom, and as O. Nigel Bolland highlights, '[f]reedom of movement was vital, for its symbolic value and also because former slaves sought to be reunited with family members and friends' (Bolland 2001, 25). Verene Shepherd stresses a gendered dimension to this negotiation by arguing that mobility was a key means through which low income women expressed agency in the post-slavery Anglophone Caribbean (Shepherd 2007, 157). Recourse to mobility could equally be a strategy for a degree of economic empowerment enabling the formerly enslaved to better establish land-based roots. Immobility and stasis were thus significant in planter attempts to control labour as well as in positive relationships between the formerly enslaved and the land. Ultimately the struggle to control mobility in the post-emancipation Caribbean was bound up in conflicting definitions of freedom held by different social groups. While it should be recognised that the mobility of labouring bodies was a singularly important dynamic in the post-emancipation era, my focus here is with external (im)mobilities and flows of information – specifically steamship mail service interactions with the region during this period. As illustrated at the start of this article, this form of mobility also shaped island histories. The focus here on Caribbean maritime mobilities forms part of a necessary consideration of historical, and also 'non-metropolitan practices' (Vannini 2011, 297).

The RMSPC's steamers entered full service in 1842, carrying post, people and high-value goods across the Atlantic and around the Caribbean archipelago. In organising the schedule, the logistics of timetabling departures and arrivals across an extensive Atlantic network necessitated a large measure of negotiation to mesh the needs of this large-scale transportation infrastructure with everyday rhythms at the various ports of call; the desired 'regularity' of the steamship service had to be modified in line with the habitual rhythms of Caribbean islands. Thus this paper is concerned both with a transimperial network that had globalising tendencies and the 'local' of its ports of call. This is entirely in keeping with the fact that, as Jonathan Rigg suggests, 'just at the time when globalisation has become the defining process of the age [...] there has emerged a vibrant concern for the minutiae and distinctiveness of the "everyday" and, by association, the local' (Rigg 2007, 10–11). Tim Edensor (2010, 1) draws on Lefebvre to underscore the significance of rhythms for understandings of place. For coastal and port town spaces, maritime rhythms of course have a particular import. The rhythms of the RMSPC's service in the Caribbean were negotiated through the local and the everyday, a scale at which, as Edensor stresses, 'regulatory processes pervade and are resisted or ignored' (Edensor 2010, 2). The incorporation of and resistance to steamship rhythms at the local scale becomes apparent through a focus on 'the banal moments of travel' (Watts 2008, 713).

Smoothing the Steamship Timetable

During the 1830s, MacQueen wrote of the predictability of steamship rhythms as a great attraction of the proposed new service. Although packet boats carried mail between Britain and the West Indies prior to the establishment of the RMSPC, this pre-existing operation was subject to disruptions (MacQueen 1838, 43). Concerning the government packet system, MacQueen lamented, 'Every thing at present is in a state of uncertainty and confusion. The sailing packets arriving at Barbadoes in unequal times and the Government steamers here being of unequal powers there is no dependence on the time of their arrival or departure at any one place and consequently neither the merchants nor passengers know how to regulate themselves' (NMM RMS 7/1, 15 March 1841). Advocating for the RMSPC, MacQueen stressed that steamship connection with the Caribbean would constitute a marked improvement upon the rhythms of government ships. Such claims tied into the 'science of energy' in the nineteenth century (Smith 1998). Crosbie Smith illustrates how networks of scientists and engineers worked to build the 'public credibility' of steam engines (Smith 1998, 5) and in contrast to the 'uneven' rhythms of sailing packets – perceived as being unfortunately tied to unpredictable natural rhythms – the RMSPC's steamers apparently promised journeys of consistent duration. Tim Edensor notes that 'Journeys have a particular rhythmic shape' (Edensor 2010, 6). In this case, the rhythmic consistency of steamship journeys was highly desirable in that it would allow for smoother trading and faster communications with the colonies (MacQueen 1838). MacQueen was able to make such claims for the new service on the basis of the public trust in steam technology carefully built by scientists and engineers.

As these new 'regular' rhythms were deemed to be a distinct advantage of steamship communications, considerable work went into constructing the steamship timetable, or 'scheme of routes'. The scheme of routes, as a series of expectations, presented the *measure* rather than the rhythms of steamship operations (Lefebvre 2004, 8). For Lefebvre, measure is 'law, calculated and expected obligation, a project'; the scheme of routes was both an ambitious and carefully calculated project (Lefebvre 2004, 8). In 1840, armed with letters of introduction from the Foreign Office, MacQueen travelled to the West Indies to investigate the requisite logistics of the operation (NMM RMS 7/1, 31 October 1840). On the basis of this trip and extensive negotiations with the Admiralty, a scheme of routes was drawn up, albeit one which demanded revisions and adaptations. By 1843, the Company had a transatlantic steamer running from Southampton to Barbados and Grenada, via Madeira. There were also various branch routes travelling from Grenada, St Thomas and Jamaica across the Caribbean (NMM RMS 36/3). Throughout the decades, the scheme of routes was re-negotiated as the printed ordering of the service played out against habits, events and needs at vessels' ports of call. Following Lefebvre, we can recognise the steamship timetable as constituting a series of 'impersonal *laws*', which were modified by '*actors*, ideas, realities' on the ground in the Caribbean (Lefebvre 2004, 6. Emphasis in original).

The RMSPC's early service offered two transatlantic passages from Southampton a month. By 1845, the Company was advertising vessels departing on the 2nd and 17th of each month 'taking a limited quantity of goods, for the following places: – Barbadoes, Demerara, Grenada, Trinidad, Jamaica, St. Vincent, St. Thomas, and Bermuda' (*The Times*, 25 February 1845). However, early on during its service, the RMSPC sought to respond to a slight inconsistency in its timetabled journeys. It was a source of concern that there were uneven monthly 'measures' in the early timetable. Thus, the Company lamented that 'owing to the inequality in time which the dispatch of the mails from England on the 1st and 15th of each month creates; and that owing to there being only 28 days in the month of February the inequality adverted to which leaves but 14 days between the arrival in course of the mails is continued as regards the mail of the 16 February, 1 March and 15 March successively. This brings these mails to Barbados *one day earlier than at other times*' (NMM RMS 6/1, 15. Emphasis added by author). To overcome this problem of journeys of uneven duration, Captain Chapman was instructed to ensure speedy fuelling at the island of St Thomas and minimal stoppage time at the intercolonial ports of call, allowing the RMS *Actaeon* to return to Barbados in a timely manner in February and March. This very specific RMSPC concern indicates the Company's broader ambitions to achieve even rhythms in its timetable to the greatest possible extent.

The presence of steamship operations in the nineteenth century Caribbean created service hubs, particularly at St Thomas, and later Barbados, which were key sites of departure for the RMSPC's branch routes through the colonies. Furthermore, the routine connections provided between colonies facilitated patterns of migration. Maritime migration was one of several mobile strategies adopted by the formerly enslaved in the negotiation of their freedom (Richardson 1985), and in this respect, steamship mobilities were significant. As well as setting people in motion, steamship operations tied people to place, particularly through coaling operations. The steamship coaling workforce was predominantly composed of women, and the RMSPC relied on immobilising strategies to secure a reliable labour source to keep their steamships moving in accordance with the timetable. Thus, coaling created a crucial intersection between steamship rhythms and the 'corporeal rhythms' of daily work (Mels 2004, 6).

Given the complications involved in seeking to deliver even and predictable journeys and the timely delivery of the mail, numerous 'modifications respecting the route of the Packets' were necessary before a workable timetable was stabilised (NMM RMS 7/1, 1 December 1840). Furthermore, any stability achieved was always only temporary, as the RMSPC's scheme of routes was constantly re-worked. Thus in addition to formal revisions of the 1841 timetable in 1843 and 1851, the timetables were constantly altered and corrected. A corrected version of the 1851 scheme of routes was produced in May 1858, for example, and a corrected version of the January 1864 scheme of routes was produced in November of that year (NMM RMS 36/2; NMM RMS 36/3; NMM RMS 36/4). Even from an official Company perspective, the steamship timetables (and thus the rhythms of individual journeys) were worked and re-worked.

The logistics of the service presented numerous mundane realities that mitigated against the production of such even rhythms. Amongst these challenges, the steamer's arrival at night posed particular logistical problems. Admiralty Agents, men with naval backgrounds, oversaw the delivery of the mail and reported to the Admiralty (to who the Company was accountable) on the RMSPC's performance. As one such Agent, Bellairs, complained at several islands there were 'no lights exhibited retarding the mail service and occasioning considerable risk of life' (NMM RMS 6/3, 12 May 1845). Even where light was available, harbours were not necessarily safe for navigation, for example, it was deemed unsafe for the Company's larger vessels to enter English Harbour, Antigua, at night (NMM RMS 7/2, 5 February 1846). The need for safe navigation caused friction in the steamship timetable as vessels were forced to negotiate nocturnal arrivals by slowing or pausing their journeys.

Even the logistical reality of achieving regular *daily* rhythms proved elusive. Although the Company initially estimated that the journey from Falmouth to St Thomas could be made in nineteen days and six hours, by January 1843 the time allowed had been revised to twenty-two days to 'cover contingencies', particularly those relating to the weather (NMM RMS 7/1, 9 January 1843). The kind of early teething problems experienced by the RMSPC were illustrated when the Company received news from Jamaica in March 1842 that since the RMS *Tweed* had not returned to Jamaica when timetabled to do so and 'Much time had … [passed] since any mails from England had been received at this Island', Captain Elliot of HMS *Spartan* had brought the waiting mail from St Thomas to Jamaica (NMM RMS 6/1, 46). It later came to light that the *Tweed*, so long overdue at Jamaica, had been delayed when the vessel ran short of coal (NMM RMS 6/1, 68). This instance of a missing mail steamer and the need for a non-RMS vessel to intervene was a far cry from the regular and comprehensive system of communication to which the Company aspired.

Weather, navigational challenges and local logistics all made a 'regular' service hard to deliver. When the RMS *Tay* was delayed en route to the Caribbean, it was reported that the vessel:

> made Barbados exactly at the end of 17 days, but the weather being hazy, and night coming on she lay to off, and drifted past the island during the night. In making up to it again and taking in some very indifferent coals she lost 38 hours. At Grenada she lost nearly two days owing to the coal depots not being so well prepared as it ought to have been, but chiefly because the Negro labourers refused to work at even high wages especially during the night. At St Thomas's she lost more than one day, from the same cause, and from a cause hitherto unexplained, she has lost above another day between St Thomas and Havana. (NMM RMS 7/1, 210–211)

In this instance, a combination of bad weather and contested labour arrangements slowed the steamer's progress. Tim Cresswell has outlined six elements of mobility: motive force, speed, rhythm, route, experience and friction (Cresswell 2010). It must be stressed that the rhythms of coaling labour were often significant to the workings of the overall service, causing 'friction' within steamers' trajectories. For example, on the RMS *Tay*'s journey in January 1842, the vessel remained at Barbados three and a half hours longer than scheduled; at Grenada the *Tay* waited almost a full day longer than timetabled in order for coaling to take place (NMM RMS 6/1, 21). Coal was also the cause of delays to the *Teviot* in 1842, although in this latter instance it was coal supplies rather than problems with coaling labour that slowed the vessel's progress. The vessel remained at Havana sixty-seven and a half hours longer than the timetable allowed because the Company had 'no coal in store' (NMM 6/1, 83).

Aside from these particular journeys, the frequently uneven rhythms of the RMSPC's early steamship service were evident from Admiralty Agent reports. The Agent on board the RMS *Clyde* indicated that far from adhering to the printed timetable, the vessel's movements in 1842 were characterised by delays. Thus at Antigua where one and a half hours were allowed for the delivery of the mails according to the scheme, three hours and fifty-seven minutes were instead 'required for this purpose' (NMM 6/1, 25). At St Vincent, where two hours were allowed by the timetable, the Admiralty Agent noted that he 'stayed five hours after the mails were on board' (NMM 6/1, 25). At Grenada, the mail was 'further detained in consequence of the men again 'striking work' (NMM RMS 6/1, 25). Due to the need for coaling and coaling labour, reliance on steam power, which promised predictable rhythms and even journeys, brought its own challenges that often made the rhythms of steamship journeys as varied as those under sail. 'Friction' proved as significant to the steamship service in the mid-nineteenth century as any new-found 'velocity' (Cresswell 2010).

Conflicts between employees could prove equally disruptive to the timetable. When the RMS *Teviot* left Vera Cruz almost 15 hours behind time, the delay was attributed to a running dispute between the captain of the steamship and the Admiralty Agent. The captain set out to sea without waiting for the return of the Admiralty Agent and the mail boat, reportedly because of his determination to 'give the mail Agent *a sweat of it*' (NMM RMS 6/1, 83. Emphasis in original). This case of poor relations between the captain and the Admiralty Agent was by no means an isolated one. In 1842, the Company's secretary suggested removing Admiralty Agents from the steamers given the 'constant interruptions and delays arising from differences between the mail agents placed on board by the Lords Commissioners of the Admiralty and the Captains appointed by the Court of Directors' (NMM RMS 7/1, 29 August 1842). Admiralty Agents had a tendency to 'interfere with the ships course or the length of stoppage at any place', which altered the rhythms of individual steamship journeys (NMM RMS 7/1, 29 August 1842).

The production of smooth rhythms in the delivery of the mail demanded a high level of cooperation between Admiralty agents and individuals based on shore. Dominica was one of the many islands at which lighting was reportedly problematic. Although 'a lanthorn' was maintained on the island, this was infrequently lit to aid the steamers' arrivals (NMM RMS 7/1, 29 August 1842) Agent Bellairs recounted his experience of arriving at Dominica at 9 pm where he fired gun signals and rockets, seeking to attract the attention of those on shore. When this was met with no response, he set out in the mail boat and 'pulled in various directions to find the town, but would not find a single object to guide me, at last [he] fell in with a brig at anchor where [he] received information how to steer for the town, and arrived there 11.10 pm' (NMM RMS 7/1, 29 August 1842). Bellairs claimed that upon landing, he was told by an army officer that he had been 'seen by several of the merchants who had been watching the boat rowing about without the slightest effort to guide or assist

[him]'. According to Bellairs, even the RMSPC's agent on the island had witnessed the scene but had not attempted to help (NMM RMS 7/1, 29 August 1842). Similarly, the Admiralty Agent serving on board the RMS *Clyde* reported that on one occasion he was forced to leave the post on board a brig at Nevis because nobody on shore was willing to assist him (NMM RMS 6/3, 18 February 1845). As in more recent contexts (Vannini 2012), smooth steamship rhythms could be produced only through collaborative working and harmonious rhythms between ship and shore.

Relations between various shipboard employees – not only the captain and the Admiralty Agent – caused such delays. On one of the *Clyde's* journeys through the Caribbean, the vessel was delayed at Grenada for two hours and forty-five minutes because of 'the men refusing to get the steam up having struck work' (NMM RMS 7/1, 24 March 1842). In response, 'several' of the crew were 'discharged and sent to England' (NMM RMS 7/1, 24 March 1842). As these various journeys suggest, both the negotiation of labour between ship and shore, and relationships between employees on board altered the rhythms of steamship journeys. Thus the efficient journeys documented within the printed timetable were re-negotiated on board ship and in port into more idiosyncratic rhythms constituted by the priorities and personalities of those working on and around the steamships. Relations between members of the crew and between shipboard and shore-based labourers ensured that slow mobility was as significant to steamship travel as speed.

Steamship Pauses

Not only was the steamship service characterised by 'friction' (Cresswell 2010, 17), but this friction was actively promoted by some service users. The RMSPC's steamship journeys indicate that although those managing the steamship network were concerned to achieve even and predictable journeys, it was the steamers' pauses that were often highly prized at ports of call, particularly for those very 'merchants' to whom MacQueen referred when arguing for the service (NMM RMS 7/1, 161–162). As B.W. Higman indicates, in managing plantations 'communication with England was the vital link, for governors, traders and planters' (Higman 2005, 121). Prior to the steamship service, various packet boat systems of correspondence with the West Indies were attempted and there was also the option of sending letters on merchant ships. Merchant ships in the late eighteenth and early nineteenth century typically made a round trip to the Caribbean in over one hundred days (Higman 2005, 131). However merchant shipping was also bound up in seasonal rhythms. As noted by Higman, such vessels were more readily found during favourable sailing conditions and when sugar was ready for export but were harder to find during the hurricane period (Higman 2005, 129).

The packet boat system that preceded the RMSPC drew occasional complaints and, at least in one location, 'the Kingston merchants were willing to use their influence to delay the departure of the packet boat when it suited' (Higman 2005, 130). These patterns continued into the era of steam. When steamers failed to stay for their full scheduled stoppage time, it frequently provoked a response. When the RMS *Avon* arrived at Bermuda from St Thomas at 6 pm on 21 February 1845 and departed twelve hours later instead of the scheduled twenty-four hours in an attempt to make up time, it provoked a 'very general complaint among the merchants and others' who had insufficient time to respond to letters (NMM RMS 6/3, 5 April 1845). These merchants in Bermuda had trading interests with other colonies in the

Americas and for those engaged in commerce, the steamers' pauses were as valued as each vessel's potential for speed. Thus, the mercantile community was preoccupied with steamship 'friction' rather than 'velocity' (Cresswell 2010, 17). 'Service delays', so easily associated with passenger anxiety in the context of contemporary travel, provided for the needs of merchants as these were better suited to their rhythms of letter writing (Bissell 2009, 66–67).

Revising the scheme of routes in 1843, the Company sought to respond to this need for suitable pauses, particularly at the larger colonies. The RMSPC explained that in the proposed timetable for 1843, 'the main object has been to remove all cause of complaint by affording to the important colonies of Jamaica, Barbadoes, Trinidad, Demerara & Berbice *longer time for replying to correspondence*' (NMM RMS 7/1, 1 April 1843. Emphasis added by author). While those managing the network stressed the potential for speed afforded by steamship technology, only significant pauses between the journeys made the service useful to those situated in the colonies.

When steamers entirely neglected given islands, the RMSPC inevitably faced a backlash from the mercantile community. The RMS *Derwent*'s failure to call at Tortola en route from St Thomas to the Windward islands in September 1850 provoked complaint from merchant John Craddock, who explained that the steamer's absence had caused him 'serious inconvenience and pecuniary loss' particularly given that there were 'few means of communication with the sister colonies other than by packet' (NMM RMS 6/6, 30 August 1850). As well as occasional instances of missing steamers, routinely curtailed stoppage time drew complaints. In May 1849, merchants from St Jago de Cuba wrote to explain their frustration with the working realities of the service. Although the printed timetable allowed for the steamer's arrival at 6 pm on the 18th of the month followed by departure at 6 am on the 20th of the month, merchants often found the packet 'sailing the same afternoon of her arrival' (NMM RMS 6/5, 22 May 1849). On the return leg of the journey, where forty-eight hours were officially allowed for answering correspondence, the merchants lamented that 'the most that is generally now conceded is 24 hours' of stoppage time (NMM RMS 6/5, 22 May 1849).

Worse than such one-off instances of steamers failing to pause on their journey were timetable changes that eliminated stoppage time altogether. The RMSPC's new timetable of 1850 allowed for only one call a month at Jacmel, with the first steamer of the month passing Jacmel on the way to Jamaica without pausing. In contrast to the RMSPC's concerns with speedy passages, residents of Jacmel desired a 'few hours' detention at Jacmel' (NMM RMS 6/6, enclosure dated 18 November 1850 in 23 July 1849). While communication between colonies was prized alongside that with European ports, the moment of the steamship's pause was crucial to allow for the preparation of communications. Whereas captains at times sought to shorten their time in port to allow them to stay on schedule or make up for delays and were concerned to deliver the 'regularity' promised by the printed timetable, the steamer's pause was particularly prized by the mercantile community in the Caribbean region. Highmore (2002, 175) stresses the ways in which 'Against the frantic circulation and accumulation of money, certain cultural practices have defined (and continue to do so) their alternative and oppositional status in relation to their slowness', however in this case it was precisely the 'accumulation of money' that prompted calls for stasis. Responses to the RMSPC's service suggest that friction was highly significant to nineteenth-century mobilities not only as a working reality of new communications technologies but also as a facet of mobility that was positively represented and valorised, in this case by elite interest groups seeking to maximise the profitability of their undertakings. Despite a rhetoric that 'place[d] greater value on high speeds', slow did not imply 'inferiority' for all groups reliant on the steamship service (Jain 2011, 1017). The case study of the RMSPC indicates that we might extend the notion of 'travel time as a gift' to encompass even requisite infrastructural delays, since in this case, the pause required to allow delivery of the mail and refuelling was beneficial to a core group of service users (Jain and Lyons 2008, 88). In this case, 'slowing down' was not simply a 'feasible compromise' for merchants, but rather, in their view, formed a necessary component of efficient communications (Vannini 2012, 103–104).

'The Great Event of the Fortnight': Steamers and Social Rhythms

Lefebvre underscores the relational significance of rhythms: 'We know that a rhythm is slow or lively only in relation to other rhythms' (Lefebvre 2004, 10). In the development of the RMSPC's service, the relationship between steamship rhythms and social rhythms at colonial ports of call proved significant. The case of Jamaica demonstrates how the relation between the steamship's rhythms and the 'sociotemporal order' in the Caribbean necessitated a negotiation, and a resolution of the two (Zerubavel 1981, xii).

Jamaica's status amongst the British Caribbean colonies ensured that Kingston was an important port of call in the RMSPC's service. The Company's 1885 timetable included a fortnightly route from Southampton to Barbados, Jacmel, Jamaica and Colon. The rhythms of the steamship service had to be re-negotiated when the timetable of 1885 brought the steamer to Kingston with inappropriate timing. Although the timetable of 1885 scheduled the transatlantic steamer to arrive in Jamaica on a Monday morning, in practice, the mail steamer usually arrived at Kingston on a Sunday. This proved problematic in that the steamers' arrival clashed with the 'cyclical ordering which organises, apportions, schedules and coordinates activities' – in this case the ordering of the Jamaican Sabbath (Edensor 2010, 8). The rhythms of this oceanic service had to be carefully intertwined with the weekly markers on shore, since Sunday was a 'most salient marker of time' and a '*milieu de mémoire*' (McCrossen 2005, 25–26). While it was understood that the extra time allowed more letters to be answered, the ship-to-shore rhythm of the delivery of the mail nevertheless provoked 'great dissatisfaction' in Kingston.

The Governor of Jamaica pointed out that 'the arrival of the mail steamer from Southampton is the great event of the fortnight and when it comes in on Sunday there is a general interruption in Kingston of the quiet and rest of the Sabbath' (TNA CO 137/522, 28 July 1885, enclosure dated 3 September 1885). Since the postal and customs departments had to work, a special train was run on the railway, and 'the cart men and drivers of public conveyances' were required to labour, the arrival of the mail steamer gave the city 'a regular week day appearance' (TNA CO 137/522, 28 July 1885). A memorial on the subject was addressed to the Governor, presented by a deputation of ministers of different religious denominations, and bearing four hundred and ninety-six signatures (TNA CO 137/522, 28 July 1885). While the memorial was led by the actions of religious ministers, these men claimed that 'even those who do not object on religious grounds still feel that it is undesirable to have a fortnightly disturbance of the ordinary Sunday quiet' (TNA CO 137/522, 28 July 1885). The petitioners here demonstrated a determination to bestow upon the seven-day week 'an aura of sanctity' (McCrossen 2005, 31). In this way, the sociocultural rhythms of Kingston – and specifically the observance of the Sabbath – were disrupted by those of the industrialised oceanic transport network. In this instance, different kinds of rhythms became inappropriately intertwined; however, the objections of colonial elites ensured that the transportation network was forced to adapt in deference to religious observance. This comprised an instance of 'overarching discourses such as those of [...] religion, or tradition' shaping the negotiation of a 'public time-space' (Mels 2004, 6).

The Anglican clergy in Jamaica had traditionally been aligned with planters' interests and had made little attempt to evangelise the enslaved population; however, Robert Stewart argues that the creation of the bishopric of Jamaica with effect from 1824 aimed to bring the enslaved into the Anglican church (Stewart 1992, 1). In contrast to the Anglican Church, Protestant nonconformist groups such as Moravians, Methodists and Baptists generally distanced themselves from the elite planter lifestyle and interacted more systematically with the black Jamaican population (Stewart 1992, 12–20). Baptists had particular political involvement, and sought to exert their influence for 'the well-being of the ex-slave' (Stewart 1992, 21). It is within this multi-denominational landscape containing groups that interacted more and less closely with the formerly enslaved that objections to steamers' Sunday arrivals were advanced. After all, as already indicated, steamship arrivals required labour to land and load mail, cargo and passengers, but were also often accompanied by the labour-intensive undertaking of coaling vessels.

When the RMSPC responded to Jamaican complaints by instructing vessels not to arrive at Jamaica until the timetabled day (Monday at 8 am), the Company was subjected to further outcry, on the grounds that this left insufficient time before the departure of the outward mail (TNA CO 137/523, 12 November 1885). While the Sunday arrival was 'objected to', the altered arrangement which brought the steamers in on schedule on Monday mornings was 'even more strongly objected to' as providing insufficient time to answer correspondence (TNA CO 137/523, 23 October 1885). Once again in this instance the mercantile, but also the wider community valorised friction in the form of steamship pauses. The RMSPC informed the colonists that their complaints would be considered when the timetable was next subject to alteration (TNA CO 137/523, 12 November 1885). The problem was rectified, but only during the course of the next mail contract. The timetable of 1890 brought the transatlantic steamer to Jamaica on a Friday, and this vessel left Jamaica on a Saturday. Unlike the 1885 timetable, the 1890 scheme of routes did not schedule any arrivals or departures in the Americas on Sundays (NMM RMS 36/4 Tables of routes for the packets of the RMSPC commencing from Southampton 9 July 1890). The Jamaican case illustrates that the rhythms of steamship services had to be negotiated and meshed with colonial norms and sociocultural values at individual ports of call. The 'weaving' of steamship rhythms into the 'urban daily reality' of Kingston (Jiron 2010, 143) necessitated a re-negotiation of an imperial project within a colonial space. Not only was there a need for 'compatibility' with 'local work routines', but maritime rhythms proved particularly thorny in that they collapsed oceanic work routines onto those of metropolitan and colonial spaces (Stein 2001, 117).

Conclusion

Steamship rhythms were negotiated with those on shore in a variety of ways. Aside from the logistical and infrastructural support required to smooth the steamship timetable, distances could only be 'annihilated' with faster communications insofar as individuals on shore were willing to cooperate to enable fast turnaround times in port or facilitate the delivery of the mail. The experiences of Admiralty Agents suggest that such ship-to-shore cooperation could not be taken for granted, and 'friction' characterised steamship operations as much as 'velocity' (Cresswell 2010). The rhythms of the service were also perceived differently in various contexts, so that the mercantile community emphasised the duration of pauses to enable efficient correspondence rather than the speed of steamship communications, while in the case of Kingston religious concerns shaped the response to the steamer's arrivals and departures and promoted arguments against velocity when it disrupted the Sabbath. The logistical, commercial and sociocultural negotiations of steamship rhythms outlined here serve to underscore how we might understand past rhythms of communication not as straightforwardly annihilating distance with speed, but rather as negotiated and balanced between the financial, social and cultural interests of groups situated at various ports of call.

The steamship service was negotiated into rhythms that were variously accelerated or slowed down: these were 'uneven rhythms' adapted to their Caribbean contexts (Highmore 2002). Furthermore, this case study indicates the fractured responses to different facets of mobility across spaces. Although from the managerial centre of the steamship network, velocity was valued and deemed to promote trade, it was rather the steamer's pause that allowed for effective commerce across the network. For merchants, the working rhythms of the service produced 'friction' not as a hindrance, but rather as an *aid* to efficiency. For those seeking to protect the Sabbath, velocity resulting in early arrivals produced a rhythmic efficiency that was equally problematic. The differentiated nature of responses to what constituted productive 'modern' rhythms in colonial port towns is perhaps suggestive for thinking forward comparatively into postcolonial urban contexts.

Acknowledgements

I would like to thank Pat Noxolo as well as the anonymous reviewers for their helpful comments and suggestions. Thanks also to David Lambert and Nigel Rigby for their ongoing support.

References

National Maritime Museum (NMM), Greenwich, London.
Royal Mail Steam Packet Company collections (RMS).
NMM RMS 6/1. Letterbook: In-letters from Public Departments 1842.
NMM RMS 6/3. Letterbook: In-letters from Public Departments 1844–45.
NMM RMS 6/5. Letterbook: In-letters from Public Departments 1848–49
NMM RMS 6/6. Letterbook: In-letters from Public Departments 1850–51.
NMM RMS 7/1. Letterbook: Out-letters to Public Departments 1839–44.
NMM RMS 36/1. First book of Routes West India Mail Service 1842.
NMM RMS 36/2. First and Second Book of Routes West India Mail Service 1843.
NMM RMS 36/3. Table of Routes West India Mail Service 1843–60.
NMM RMS 36/4. Table of Routes West India Mail Service 1860–1903.
The National Archives, UK (TNA).
TNA CO 137/522. Colonial Office Jamaica, Original Correspondence 1885.
TNA CO 137/523. Colonial Office Jamaica, Original Correspondence 1885.

Bissell, D. 2009. "Visualising Everyday Geographies: Practices of Vision through Travel Time." *Transactions of the Institute of British Geographers* 34 (1): 42–60.
Bolland, O. N. 2001. *The Politics of Labour in the British Caribbean: The Social Origins of Authoritarianism and Democracy in the Labour Movement*. Kingston: Ian Randle.
Cresswell, T. 2006. *On the Move: Mobility in the Modern Western World*. New York: Routledge.
Cresswell, T. 2010. "Towards a Politics of Mobility." *Environment and Planning D: Society and Space* 28 (1): 17–31.
Cresswell, T. 2011. "Towards a Politics of Mobility." In *African Cities Reader: Mobilities and Fixtures*, edited by N. Edjabe and E. Pieterse, 159–171. Vlaeberg: Chimurenga and the African Centre for Cities.
Edensor, T., ed. 2010. *Geographies of Rhythm: Nature, Place, Mobilities and Bodies*. Farnham: Ashgate.
Eriksen, T. 2007. *Globalization: The Key Concepts*. Oxford: Berg.
Geggus, D. 1985. "Haiti and the Abolitionists: Opinion, Propaganda and International Politics in Britain and France, 1804–1838." In *Abolition and Its Aftermath: The Historical Context, 1790–1916*, edited by D. Richardson, 113–140. Oxford: Frank Cass.
Hall, D. 1971. *Five of the Leewards 1834–1870: The Major Problems of the Post-emancipation Period in Antigua, Barbuda, Montserrat, Nevis and St Kitts*. Barbados: Caribbean Universities Press.
Hall, N. 1992. *Slave Society in the Danish West Indies: St Thomas, St John, and St Croix*. Baltimore, MD: John Hopkins University Press.
Hall, D. 1993. "The Flight from the Estates Reconsidered: The British West Indies, 1838–42." In *Caribbean Freedom: Society and Economy from Emancipation to the Present*, edited by H. Beckles and V. Shepherd, 55–63. Kingston: Ian Randle.
Headrick, D. 1981. *The Tools of Empire: Technology and European Imperialism in the Nineteenth Century*. New York: Oxford University Press.
Highmore, B. 2002. "Street Life in London: Towards a Rhythmanalysis of London in the Late Nineteenth Century." *New Formations* 47 (2002): 171–193.
Higman, B. W. 2005. *Plantation Jamaica 1750–1850: Capital and Control in a Colonial Economy*. Kingston: University of the West Indies Press.
Jain, J. 2011. "The Classy Coach Commute." *Journal of Transport Geography* 19: 1017–1022.
Jain, J., and G. Lyons. 2008. "The Gift of Travel Time." *Journal of Transport Geography* 16: 81–89.
Jiron, P. 2010. "Repetition and Difference: Rhythms and Mobile Place-making in Santiago De Chile." In *Geographies of Rhythm: Nature, Place, Mobilities and Bodies*, edited by T. Edensor, 129–143. Farnham: Ashgate.
Lambert, D. 2008. "The 'Glasgow King of Billingsgate': James MacQueen and an Atlantic Proslavery Network." *Slavery & Abolition* 29 (4): 389–413.
Lefebvre, H. 2004. *Rhythamanalysis: Space, Time and Everyday Life*. London: Continuum.
MacQueen, J. 1838. *A General Plan for a Mail Communication by Steam between Great Britain and the Eastern and Western Parts of the World*. London: B. Fellowes.
McCrossen, A. 2005. "Sunday: Marker of Time, Setting for Memory." *Time & Society* 14 (1): 25–38.
Mels, T. 2004. "Lineages of a Geography of Rhythms." In *Reanimating Places: A Geography of Rhythms*, edited by T. Mels, 3–42. Aldershot: Ashgate.
Peters, K. 2012. "Manipulating Material Hydro-worlds: Rethinking Human and More-than Human Relationality through Offshore Radio Piracy." *Environment and Planning A* 44 (5): 1241–1254.
Richardson, B. C. 1985. *Panama Money in Barbados 1900–1920*. Knoxville: The University of Tennessee Press.
Rigg, J. 2007. *An Everyday Geography of the Global South*. London: Routledge.
Schivelbusch, W. 1977. *The Railway Journey: Trains and Travel in the 19th Century*. Oxford: Blackwell.
Shepherd, V. 2007. *I Want to Disturb My Neighbour: Lectures on Slavery, Emancipation and Postcolonial Jamaica*. Kingston: Ian Randle.
Smith, C. 1998. *The Science of Energy. A Cultural History of Energy Physics in Victorian Britain*. London: Athlone Press.
Stein, J. 2001. "Reflections on Time, Time-space Compression and Technology in the Nineteenth Century." In *Timespace: Geographies of Temporality*, edited by J. May and N. Thrift, 106–119. London: Routledge.
Stewart, R. J. 1992. *Religion and Society in Post-emancipation Jamaica*. Knoxville: The University of Tennessee Press.

Vannini, P. 2011. "Mind the Gap: The *Tempo Rubato* of Dwelling in Lineups." *Mobilities* 6 (2): 273–299.
Vannini, P. 2012. *Ferry Tales: Mobility, Place, and Time on Canada's West Coast*. New York: Routledge.
Watts, L. 2008. "The Art and Craft of Train Travel." *Social & Cultural Geography* 9 (6): 711–726.
Zerubavel, E. 1981. *Hidden Rhythms: Schedules and Calendars in Social Life*. Chicago: University of Chicago Press.

Learning 'Large Ideas' Overseas: Discipline, (im)mobility and Political Lives in the Royal Indian Navy Mutiny

ANDREW D. DAVIES

Department of Geography and Planning, University of Liverpool, Liverpool, UK

ABSTRACT *In February 1946, the 20,000 sailors of the Royal Indian Navy, the colonial navy of the Government of India, mutinied. Having a number of grievances, from colonial rule of India, inefficient demobilisation procedures and ill treatment from superior officers, sailors on ships and shore establishments across the Indian Ocean took part in the mutiny, which represented the largest time a military force had disobeyed British Rule since the Mutiny of 1857. This paper examines the ways in which the geographies and mobilities of naval service shaped the political lives of the sailors in the RIN. On the one hand, both military (naval) and colonial forms of discipline worked through the spaces of the ship to attempt to control and order sailors' lives. On the other, the mobile nature of life at sea, travelling from place to place and encountering colonial difference within the RIN, opened the door to different political ideas to influencing the sailors. At the same time, far from being a disconnected space, separate from the land, the naval ship combined with sailors' land-based connections allowed them to contest and rework 'landed' political activity from the sea.*

Introduction

In the aftermath of the Second World War, British India experienced what has been classified as a nationalist 'upsurge' (Chandra et al. 1989) against colonial rule. Numerous outbreaks of disorder occurred across India and the pace of decolonisation increased. One of these events was the mutiny of the Royal Indian Navy in February 1946.[1] Starting on the 18th of February in Bombay, ratings and non-commissioned officers[2] of the RIN refused to report for duty, and in some cases eat. Using the telegraphy skills they gathered in the Navy, the sailors began spreading news of their refusal to work across the Indian Ocean. Combined with widespread press reports of events in Bombay, fellow sailors in other naval ships and establishments as far away as Aden and the Andaman Islands began to mutiny. Over the next week, tensions mounted between the sailors and the colonial authorities. Ratings marched through

Bombay, and were accompanied by civil disturbances. In Karachi, sailors on board an old sloop, HMIS *Hindustan,* exchanged fire with soldiers ashore. The imperial authorities responded by threatening the sailors with destruction. Admiral John Godfrey, commander of the RIN, read a statement out via all India Radio on the 21st of February which told the ratings in no uncertain terms that superior military forces were prepared to attack them if violence persisted (ADM 1/19411). Following a day of tension, the ratings returned to duty in Bombay on the 23rd, and were followed by those in other bases over the next few days.

Whilst dwarfed by the later violence of communal rioting in Calcutta in August 1946 and the partition of India and Pakistan in 1947, the RIN mutiny is an important, yet often overlooked, moment in the history of decolonisation. Whilst there had been small mutinies amongst allied forces after the end of Second World War, the RIN mutiny was the first time that virtually the whole of a colonial military service had rebelled since the 1857 mutiny/rebellion in India. Around 20,000 sailors refused duty, either violently or in non-violent solidarity. News of the mutiny spread quickly around the world and the colonial authorities were acutely concerned about the threat to their continued ability to govern India, setting up a Commission of Inquiry which ran between April and June, 1946.

The mutiny itself had many causes, from nationalist anti-colonialism amongst some sailors through to more mundane grievances raised by the sailors such as poor treatment and conditions during service, and to search for one 'key' cause is somewhat misplaced (Davies 2013a). However, this paper focuses not on the causes of the mutiny per se, but rather on the ways in which the doctrines and lived experiences of life in the RIN, particularly the spaces of the ship, but also life ashore, functioned as social arenas in which the consequences of colonialism played out in distinct ways.

This paper draws upon textual sources from a variety of locations to uncover how mobility formed a key aspect of the RINs attempts to structure and control its sailors, and how in turn, the sailors resisted these disciplinary practices. These include, primarily, the official Reports of the Commission of Enquiry into the RIN Mutiny of 1946, the written transcripts of the oral testimonies given at the Inquiry and the written submissions of witnesses given to the Inquiry beforehand. These sources, together with a number of other archival sources were located in the National Archives of India, New Delhi. Other sources were drawn from the India Office Records collection at the British Library in London, and Admiralty records in the National Archives in Kew, London. Finally, a number of sailors involved in the RIN, both officers and men, have published their own accounts of life in the RIN in the 1940s, which have also been drawn upon in places. Examining these statements allows us to uncover the ways in which discourses and disciplinary practices were played out and resisted by the men affected by them.

The paper begins by giving a background to the Royal Indian Navy and the events of February 1946. It then moves on to situate the paper in relation to discussions about the geography of ships and the sea more generally. Two, later, more empirical sections focus on the practices of discipline and resistance within the RIN. Through these sections, the intersections between temporality and mobility, particularly spatial confinement aboard ship and circulation through wider society show how the mobility of the sailors of the RIN was important to shaping their political outlook. The paper concludes by reflecting on how shipboard life in the RIN is important to not only our understandings of the 'geography of the ship', but also how historical understandings of mobility are equally as important as more contemporary studies.

The Royal Indian Navy

Military forces around the coast of India were irregular throughout the colonial era. Although the RIN could trace its lineage back to the Bombay Marine in the seventeenth Century, colonial naval forces were in British India notably inferior to the Royal Navy (RN). The RIN itself was formally established in 1934, from the smaller Royal Indian Marine. The RIN remained small and ill-equipped until the outbreak of Second World War, when it increased in size, particularly after the Japanese declared war on Britain in December 1941. The RIN was well placed to assist the RN and help to escort vessels across the Indian Ocean, but its ships served in the Mediterranean, Atlantic and Pacific too. As a result, from a force of 1451 ratings in December 1939, the RIN employed 22,291 ratings in 1945 (Report of the Commission of Inquiry). This massive increase in manpower laid the foundations for what was to become the Indian Navy, but also put a considerable burden on the administration of the RIN. New ships and shore installations were built, and recruitment officers struggled to attract the required number of potential sailors to meet the requirements of the war. This led to a host of issues, as recruits were told a variety of exaggerations about their lives in the navy, from being guaranteed jobs for life, to not being informed that they may be asked to perform culturally taboo tasks like cleaning the ship's toilets. The increase also meant a lack of experienced sailors to train and manage new recruits in ways that were deemed appropriate. As a result, bullying, verbal and racial discrimination were common across the RIN – to take only one instance from many, Leading Telegraphist Akram in HMIS *Talwar* reported to the Commission of Inquiry that Muslim sailors who were praying on board HMIS *Bahadur* were kicked by a 'Gunner Smith' as they hadn't stood up to attention when he entered the room to carry out the daily inspection (Bombay Witnesses, 127). Having served in the war, many sailors were then demobilised as both the British and Indian authorities rationalised military spending. In many cases, this meant being placed in an overcrowded shore establishment with other sailors, before being formally discharged from the service. It is not without some merit that Madsen (2001, 303) called service in the RIN a 'prolonged ordeal' for its sailors.

These tensions and grievances came to a head in the Mutiny of 1946. On the 17th of February, the ratings of HMIS *Talwar*, the RIN signal school[3] located in the Colaba district of Bombay, refused to eat the food that was served to them at breakfast, and thereby disobeyed a direct order. The standard of food served in the RIN had been poor for some time: an official report conducted six weeks before had already pointed out how supplies of staple foods were not fit for animal consumption (Report of the Commission of Inquiry). However, in *Talwar* this was the latest incident in a recent history of tensions between sailors and the authorities. The establishment itself was overcrowded with men awaiting demobilisation following the end of Second World War and had approximately 1000 men stationed in it (Hastings 1988). On the 2nd of February, BC Dutt, a sailor based in *Talwar,* was arrested painting seditious slogans on the walls the day before the Commander in Chief of British India arrived for an inspection of the base. Dutt in particular was a radicalised sailor who had grown to detest the discrimination he faced in the Navy. However, given the wave of nationalist sentiment occurring across India, the RIN found itself affected by increased awareness of the struggle for Indian Independence amongst its sailors. Then, on 8th of February, the British Commanding Officer of *Talwar*, Commander King, rebuked some of his men for making catcalls at members of the Women's Royal Indian Navy (WRINs).[4] Sailors later alleged that King had sworn at the men, calling them variously 'black bastards', 'coolies' or 'junglies' (Dutt 1971; Das 1994).

Events in *Talwar* proved the starting point for what became a week of confused agitation spreading across the Indian Ocean. Sailors in different ports and aboard different ships reacted in different ways. Some, like Dutt, saw the opportunity for a violent revolution to remove the British from India. Others used the fraternal bonds they had forged through their military service to go on sympathy strike (for example, in HMIS *Hooghly* in Calcutta, ratings presented a list of issues they wanted to address, one of which was the release of BC Dutt from imprisonment [ADM 1/19411]). In Bombay *Talwar* and Castle Barracks became centres for popular protest throughout the city, mutinous sailors marched through the city and riots broke out. Likewise, in Karachi, sailors heard about events in Bombay on the 19th, and began marching through the city on the 20th (Deshpande 1989). As the colonial authorities struggled to regain control, fighting broke out between the RIN sailors and loyalist soldiers and police in Bombay and Karachi. Bombers were flown over Bombay harbour to intimidate the sailors, whilst larger and better equipped RN ships were called to the area. Given the increasing odds being ranged against them, together with a lack of support from nationalist political figures like Nehru and Jinnah meant that the sailors in Bombay surrendered early in the morning on the 23rd.

The RIN mutiny combined a number of different political issues, from the increasing speed of decolonisation, the poor organisation of the RIN, differing forms of nationalism (militant, communist and elitist), which all combined in the violence in Bombay and Karachi. However, life within the RIN mean that unlike other events in India in 1946, there was a maritime nature to the mutiny, and that this has important implications for our understandings of how discipline played out in liminal areas of colonial rule.

The Sea and the Ship

At first glance, it is deceptively easy to think of 'the ship' as a space of singularity. Existing as a material object which travels, it seems in many ways to be an 'immutable mobile', or a distinct spatial technology that, through its movement, can make and maintain networks between other places (Law 1986; Law and Mol 2001). Whilst travelling at sea, it is also tempting to classify the ship as an almost carceral space, with those on board being cut off from the land and the rest of human society. However, both of these readings simplify the diverse social and corporeal effects that 'being' on board a ship can produce, which in this paper are implicated with colonial systems of governance. However, the relational nature of life at sea means that ships are also striking places of encounter, as many geographers have argued in recent times (Ogborn 2002; Driver 2006; Ogborn 2008; Hasty 2011; Hasty and Peters 2012). Key here is the need to recognise that the 'ship' as any other space is relationally produced, and therefore different ships are productive of different subjectivities – see for example Ryan's (2006) account of the making of the ship as a domestic space. However, as mobile spaces of potentially global travel, ships are also translocal spaces (Conradson and Mckay 2007) which are in turn implicated in the numerous subjectivities and corporeal relations that they encounter, and are of clear importance to mobility scholars (Gogia 2006). Again, this bears witness to ships' seemingly paradoxical ability to both contain and discipline those who live and work aboard them, yet also to produce more diverse, cosmopolitan identities (Davies 2013b; Ong, Minca, and Felder forthcoming). Yet, as Adey (2006) has pointed out, even immobility is relative, and is always embedded within a wider political, cultural and social milieu.

The perceived paradox between freedom and control within the naval (or indeed any other) ship forms part of wider social and cultural readings of the sea. In *A Thousand Plateaus* (1987), Deleuze and Guattari apply their conception of the smooth and the striated to the sea. On the one hand, the sea is the ultimate 'smooth' space, affecting and being affected with little sense of order. It is from here that the sense of the sea as mysterious, unknowable and, potentially, as a space of freedom emerges – the sea is beyond the control of governments or other forms of authority, and escaping onto the high seas provides a route into freedom for those who seek it. The material nature of the sea, existing as ice, water, salt, spray, fog and cloud makes it hard to classify – it is alive with possibility. On the other hand, the sea is highly interdigitated with systems of international law and control – it is 'striated' and ordered by human authority. States in particular can attempt to inscribe their power onto the sea – navies enforce military control and can blockade certain parts of it, territorial waters can be marked upon it, ships travel along particular 'sea lanes' on it, and international laws govern just what is allowed to take place on-board ships (see, for example, the chapter on ports and international law in Nordstrom 2007). Thus, taken in a Deleuzo-Guattarian sense the sea and the social intermingle, producing a watery assemblage of materials, organisms and discourses, that ebb and flow between the (rhizomatic) freedom of the smooth and the (arborescent) control of the striated.

Thus, ships occupy seaspaces characterised by the limits to their mobility. However, the sea and the ship are also productive of specific types of mobilities. Whilst 'mobilities' has tended to focus on contemporary forms and spaces of travel/movement, work on historical mobilities is becoming increasingly visible, especially work on ships. For example, Ashmore (2013) has recently argued that the 'slowing down' of mobility caused by sea travel, where journeys can take weeks or months, offers a new way to explore how 'passengering' is a distinct societal process. Travelling long distances at sea is undoubtedly a slower and more temporally extended type of transport than many contemporary forms. Whilst it has long been recognised that there is a relational politics to (im)mobility (Adey 2006; Cresswell 2010), the naval ship, especially in the past, operates as a space precisely because it is (relatively) slow moving and removed from the land. As a result, sea travel is also shaped by technological and social systems which facilitate these extended periods at sea (Law 1986). This results in naval ships becoming spaces where practices of discipline and order are seen as paramount. On the one hand, this can result in the popular ideal of the moral hierarchy of shipboard order, embodied in the notion of the captain holding ultimate responsibility for the safety of the ship (Foucault 2007). However, ships also form part of wider networks of societal control and ordering, as Ong, Minca, and Felder (forthcoming) have recently argued in their study of the biopolitical control of bodies as they crossed national borders, the infrastructure of shipping lines formed a key socio-technical aspect of managing and disciplining populations.

This movement between order and disorder aboard ships has also been at the heart of wider representations of those who live work on them. By travelling at sea, being far from 'home', not even necessarily of any one ship, the sailor is characterised as being less constrained by conventional norms and behaviours. However, this liminality is not uniform, and can be imagined stretching along a spectrum from the conventional 'jack tar' who is full of fun, through to the smuggler who transports elicit goods (Hyslop 2009), to the dangerous and sometimes revolutionary image of the mutineer (Featherstone 2009). Land (2006) exemplifies the ways in which difference is encountered and mobilised through the body of the sailor. Examining colonial disciplinary practices, Land argues that sailors occupied spaces that belonging neither to the metropole nor to the periphery, and had to perform a number of negotiations with the authorities in order to claim that they were 'civilised' and should therefore be exempt from corporal forms of punishment.

Thus, both the representation and lived experience of being 'a sailor' are shaped not only by the practices of life at sea, but also by the imagined geographies and potential of maritime life as it is viewed from the shore. These are often productive of stereotypical imaginings of sailors as liminal which were mentioned above. However, of specific interest to this paper, the naval sailor is also marked by the militarism inherent in his work, in particular through the naval organisation's attempts to enforce discipline and control through restricting or routinising of his daily practices. The containment and order of life aboard a naval ship is seen as fundamental to existence as a sailor, with its required understanding of specific technical and terminological knowledges necessary to work in a military organisation at sea. Thus, whilst ashore, the naval sailor is often represented as an unknown other, at sea he is relied upon to act in an orderly and trustworthy fashion. Indeed, this distinction between a sailor whilst ashore and a sailor at sea showcases the mutable and unsettled nature of the sailor as an imagined category. Foucault exemplifies this when he classified the ship as a heterotopia,

> a floating piece of space, a place without a place, that exists by itself, that is closed in on itself and at the same time is given over to the infinity of the sea and that, from port to port, from tack to tack, from brothel to brothel, it goes as far as the colonies in search of the most precious treasures they conceal in their gardens. (Foucault 1986, 27)

Key to Foucault's metaphor of the ship here is his reliance upon the imagined geographies of the ship, and by extension its crew, as a type of immutable mobile, travelling from brothel to brothel, colony to colony, to argue for a society of openness and adventure. The boat is a specific place, whilst the sea it sails on/in is intensely placeless and empty. The ship's mobility is of crucial importance here – as the ship travels from place to place, it is altered by the experience of travelling, but there is a clear lack of engagement with the role of travelling in these heterotopias here – it is the destination, or the point of origin that are important to Foucault, and the journey or experience of travel across the space remains peripheral to the idealised notion of the ship as a distinct space of freedom. This lack has begun to be addressed in studies of maritime mobilities as discussed above, but it is important in perceptions of the naval ship as a governable, controlled space. This is especially manifest in the socio-material relations of a ship whilst at sea, and a ship when it is moored in harbour. For example, Dening (1992) in his study of the mutiny on HMS *Bounty* in 1789 has argued that the *Bounty* was a manageable space whilst at sea, but once in harbour in Tahiti the sailors became disgruntled and rose up in the next stage in their voyage as a result of the relations between the ship's captain, the crew and the peoples of Tahiti. It is the liminal, fluid and contested relations between different people within this social context that eventually broke down the naval discipline of HMS *Bounty* and produced the infamous mutiny.

In the case of the RIN Mutiny, the contested causes of the mutiny ranged from an aggressive anti-colonialism through to less formally 'political' grievances about the poor standard of food served to the sailors. The lack of consensus is important as it shows that despite the best efforts of the RIN's organisation to manage and control the lives of the sailors under its command (and the acceptance of these disciplinary practices by many sailors), this effort was undermined by both the colonial discourses that underscored its doctrines and the resistant actions of the sailors themselves. In this case, nationalist thought was created and fomented in particularly naval spaces which were often overseas, yet were connected to the homeland of India. The varied and contingent connections to the mainland were interpreted and utilised by sailors whose ideas about home, independence and the nature of colonialism were actually constructed and shaped by being a part of naval, maritime life. The next section of the paper begins to deal in more depth with the actual practices of control and order in the colonial RIN, specifically dealing with the lived experiences of various types of mobility on a naval ship.

Rhythm and Discipline in the Colonial Navy

The naval ship of the 1940s is a uniquely relational socio-material assemblage. Whilst seeming like a stable, coherent object, the ship itself is composed of many components. The material structure of the ship's hull and superstructure was combined with the many components, such as engines, guns, ammunition and more complex equipment like radar and sonar (or ASDIC, in 1940s British naval nomenclature). However, many of the objects that make up the ship were not permanently attached to the ship. Wear and tear or battle damage to a ship's machinery meant that repairs had to take place, and outdated technological equipment had to be replaced, especially as Second World War drove technological advancement at a rapid pace. The crew of the ship are also inherently a part 'of' it as they ensure the ship functions effectively, as we shall see below. Crew members, however, were not permanent fixtures, and would leave and be replaced over time. As a result, the material and social structure of the ship changed over time. Ships could be sunk in combat, but even undamaged ships would decay over time as component parts were weathered at sea and worn down through use until they were replaced, or the ship itself was deemed surplus to requirements and sold or broken up for scrap.

However, in the case of the RIN itself, its ships were conditioned by its position as the naval force of a colonised nation, or the 'unwelcome stepchild' of the RIN (Madsen 2001, 300). This meant that the RIN was not a large naval force, and was composed mainly of small escort-type vessels, intended to help with convoys or to conduct coastal activity. It was also paid for from the treasury of the Government of India. Whilst there had been debates about expanding the size of the Navy to include larger ships that would allow the RIN to project a degree of geopolitical power for a future independent India (as the RN traditionally did for Britain), the end of Second World War and the uncertainty surrounding the future of India as a colony meant that these plans were on hold (Hastings 1988). As a result, the RIN in 1946 was composed of whatever ships and equipment the Government of India had been able to pay for and negotiate from Britain during the war.

As an explicitly military object, the RIN ship's main purpose, beyond 'projecting' power, was to ensure military victory at sea, and at the most brutal level this meant that the purpose of a naval ship was to make sure that when a weapon was fired, the enemy was damaged or killed as a result. To ensure that the assemblage of the ship was an effective object able to meet military goals, specific doctrines are developed in each navy which determine the lives of sailors on board ships. Naval doctrine on board ship shaped the life of naval sailors. In the first instance, this was about conditioning men to suit military life. This process is well documented, and often takes the form of an attempt to remove individuals in the military from their civilian identities and to rework loyalty to ones new comrades in the military (Corona and Godart 2010). This removal from 'civilian' life was never likely to be 100% successful as individual sailors negotiated their own positionality within these doctrines, but represents a common theme throughout military history (Bearman 1991). However, in a colonial navy, these doctrines also served to help govern and manage the colonial subject. Firstly, containing and maintaining a healthy and disciplined body was always of concern to military and colonial authorities (Land 2006; Wald 2012). This, for the RIN, meant creating a uniform body of sailors who could be treated in similar ways (Davies 2013b; Spence 2014). The RIN encouraged men to forget communal identities (such as a specific Hindu or Muslim identity) and to mix with other sailors equally. When asked under testimony, Chief of Naval Personnel, Commander James Jefford stated that this policy of mixing men together was amongst the 'best things' the Navy had done (Delhi Proceedings, Bombay Witnesses, 70). Thus, entering colonial military service marked a transition to a new military life where civilian identities were to be left behind. Military life in the RIN was also inevitably structured by colonial ideas about race and superiority. Official booklets given to officers to aid them in understanding the variety of Indian cultures they could encounter were paternalistic and relied upon deterministic reasoning, stating freely that '[t]he characteristics of the peoples vary according to their geographical and climatic surroundings' (Creeds and Customs of the RIN, 6). Likewise, Commodore Lawrence, the Chief of Staff at Naval HQ in Delhi gave testimony to the Commission of Inquiry in May 1946 that Indian sailors took longer to learn things than RN sailors (Bombay Witness statements, Vol. 2).

The second aspect of doctrine was the governing of sailors' behaviour whilst on board ship, and the subsequent embodiment of naval doctrine in the sailor. Even as far back as the seventeenth century, the Royal Naval ship was an ordered space where the sailor's life was open to inspection, as Dening (1992) has argued in his classic account of naval life and discipline, and the subject of the naval sailor could be controlled by having his personal belongings inspected at a moment's notice. This process of managing life on board a ship was maintained in the RIN. The British naval ship is divided into particular spaces, with officer's quarters traditionally being to the rear of the ship, with the crew quartered forwards. Crew's mess decks were communal living spaces, where off duty men slept, ate and lived together. This created a distinct binary between officers and men (McKee 2002), but also created the ship as a space of limited mobility depending on an individual's rank. The organisational structure of the military ship was therefore clearly ordered by disciplinary practices of limiting spatial mobility. In some ways then, the ship can be read as a military 'camp' where bodies are contained in specific ways (Wald 2012), and clearly relative practices of (im)mobility were important markers to the rank and position one held on board (Adey 2006). In the RIN, this was complicated by issues of race and caste, where sailors were recruited and told they would not perform tasks which would affect their caste or communal identity, yet were told to do things like cleaning the toilets (Dutt 1971). Thus, the application of RN doctrine without reference to South Asian cultural norms resulted in increased tension in the service: sailors had already mutinied in nine individual ships during the war – for example, HMIS *Khyber* mutinied in the UK in September 1942, with the official cause being reported as 'refusal to work due to caste restrictions' (Report of the Commission of Inquiry, 24).

Beyond these tensions which ran across the service, there was also a distinct rhythm to naval life which informed sailors' mobility within the material space of the ship. As Cresswell (2010) has argued, rhythm is an important part to mobility as it conditions how people move and perform in particular space, but also allows actions to be measured temporally. Days at sea passed according to the strictures of keeping watch, which in the RN and RIN meant dividing the crew into two groups, the 'Port' and 'Starboard' watches, and dividing the day into five 4-h watches, and two 2-h watches. These alternated so that the different parts of the crew would not be responsible for the 'Middle' watch between midnight and 4 am on consecutive nights. This system operated continually at sea, interrupted only by a call to action stations if the enemy were encountered. Thus, life at sea became one of unceasing regularity, with a clear rhythm being established as the ship travelled. The rhythm ebbed and flowed, and sleep patterns were never constant as the two watches alternated back and forth.

This temporal rhythm provided an overall shape to life at sea, but was also impacted by the tasks that sailors were asked to undertake whilst afloat and on duty. The complex nature of the naval ship meant that the crew had to work together to make it function effectively. 'Drill' became a key part of this process – repeating the tasks one is supposed to perform so that actions become second nature under combat. These tasks necessarily created routines and orders within the wider ship, such as gun drills to ensure the ships' weaponry functioned effectively, through to exercise regimes intended to keep the crew physically fit and able to act at all times. Importantly, these practices often meant working as a team – the loading of a gun, for instance, required many bodies to move in precise, structured ways – opening and closing the gun's breech, loading the shell correctly, aiming, compensating for the movement of the ship, etc. – so that the procedure for inflicting death upon others was done as safely for the 'friendly' crew as possible. Dening (1992, 82), writing about the mutiny on HMS *Bounty* usefully describes the effect was of this routine over time.

> On a ship, as on a battlefield, every event needed to be predictable and every response instinctual. So every place, every occasion, every action had its definition and its rules. However, unlike a battlefield where the experience of the unforeseen is usually a small part of a soldier's life, on a ship, every day and night, men experienced the value of efficient, instinctual behaviour in the face of the unpredicted.

Thus, time at sea was, in any navy, crucial to ensuring that a sailor was indoctrinated correctly, especially in the RIN where different ships of different sizes and roles required different doctrines/drills. More than other branches of the colonial military, though, the RIN came to see the experience of travelling at sea as a key way to ensure that its sailors learned and inhabited these new identities. In a statement given to the Commission of Inquiry established after the 1946 Mutiny, Lieutenant-Commander AK Chatterji, an Indian Officer at HMIS *Chamak,* a radar training establishment in Karachi, remarked 'A spell at sea is very good for developing discipline and pride of the service, but due to expansion [of the service during wartime] a large number of officers and ratings had to be employed ashore in training establishments … and comparatively few got sea experience.' (Karachi Statements, National Archives of India, 143). The removal of the sailor from the land aboard a ship for an extended period of time shows the temporality inherent within these mobile practices – following Ashmore (2013) this is partially about a temporal extension of mobility over days and weeks, but one that relies upon this extension to produce distinct social relations – in this case, a military ship that, through repeated training and drill, can be relied upon to function effectively in a time of crisis.

Thus, the mobility of life on board a ship was not only one of movement across the sea's surface, but, is also about the practices of immobility, rhythm and embodied movement necessary to create an 'effective' naval ship. Thus, RIN discipline inherently utilised tactics of routinised practice and spatial discipline aboard the travelling naval ship to co-constitute the affective and embodied lives of the sailor at sea. In the views of Percey Gourgey (1996), an ex-RIN Lieutenant, 'Years of training and discipline had so conditioned [the ratings] reflexes that disobeying the orders of their superior was difficult if not impossible' (9). Clearly, however, the sailors of the RIN proved themselves able to disobey orders: not only did they mutiny in 1946 – there had been mutinies on board individual ships throughout Second World War. Colonial authorities put the cause of these mutinies down to the ratings' 'exaggerated notions about their own rights' (Report of the Commission of Inquiry, 9). The next section of the paper turns towards the ways in which the ratings of the RIN were perfectly aware of their rights, and were able to apply their knowledge to shore-based struggles as well as the injustices they faced at sea.

'Landed' Relationships and Naval Disorder

Whilst the RIN's authorities sought to control and manage the lives of the sailors under their command, their ability to contain a sailor, and therefore discipline his body was always limited. As already mentioned above, the ideal way of conditioning a sailor by sending him out to sea for an extended period of time was not always possible for the RIN. However, simply thinking that being at sea was enough to stop the sailors from mutineering is somewhat naïve. For example, whilst HMIS *Shamsher* was one of the few ships that did not mutiny, and happened to be at sea, it also had an Indian officer as its captain who talked down the sailors under his command (Written Statement of Lt. Krishnan, Witness Memo's Submitted to the Commission of Inquiry). Similarly, HMIS *Kathiawar* sailed into Bombay harbour on the 23rd of February in a state of mutiny, after the majority of sailors ashore had surrendered (ADM1/19411). Relations between officers and men on board individual ships were therefore of clear importance to sailors decisions to mutiny. However, there are also clear linkages to events ashore which need to be explored in more detail in order to understand how shipboard identities were shaped both on board, but also by connections to other, non-ship spaces.

Firstly, we must consider some of the effects of movement and education that a naval lifestyle had on many of the sailors of the RIN. Throughout much of the witness testimony given to the Commission of Enquiry, sailors and officers speak of how well educated and politically aware the sailors were. For many sailors, this education came through being mobile and travelling around the world. For instance, one key method by which sailors became radicalised through their service was by experiencing democracy overseas. The RIN had ships based in the Mediterranean and Atlantic. Having sailors based overseas in the fight against Fascist Germany, Italy and Japan meant that it was only a short step for sailors to begin to ask questions about the justice of colonial rule. The grievances that these questions raised were well known. For instance, the Muslim daily newspaper *Dawn* quoted on the 24th of February a statement by Asaf Ali in the Central Legislative Assembly 'During the war, the ratings had been hearing about freedom and national self-respect and their conduct in the war has been praised by Mr Attlee. But they found Commanding Officer King calling them names which I am ashamed to repeat' (Newspaper clippings bearing on the Mutiny). Officials were aware of the effects that travel had on fostering emergent political sentiment amongst the ratings too – when a mutiny broke out on board HMIS *Konkan* in 1942 when it was stationed in Tobermory, Scotland, the one of the official responses was that sailors 'should not remain in the UK longer than absolutely necessary as contact with RN ratings is apt to develop "large ideas" amongst RIN ratings.' (Report of the Commission of Inquiry, 21). Clearly then, naval discipline required, from the military's perspective, careful spatial and temporal management in order to maintain its effectiveness, and whilst 'mixing' amongst fellow colonial subjects was allowed as seen above, other forms of boundary crossing were to be limited. Crucially then, as subjects of a still global empire, RIN sailors found their lives being managed differently as they voyaged across the empire. Here, the embodied nature of how travel is experienced differently according to one's subjectivity is clearly important (Gogia 2006).

However, beyond simply questioning colonial rule and the many injustices which they faced as a part of their daily naval routines, there was also a more deep-seated radicalism at work. Whilst sailors travelled, particularly through the Indian Ocean, they came into contact with other struggles. RIN sailors were involved in campaigns in Indonesia as British forces assisted the Dutch in regaining their colonial possessions. However, more important for many sailors was the fact that whilst fighting in Burma they came into contact with Indian soldiers who had fought in Second World War alongside the Japanese against the British. These soldiers, many of whom had been captured in the fall of Singapore in 1942, formed the Indian National Army (INA, or Azad Hind Fauj), led by the radical exiled nationalist leader Subhas Chandra Bose. Bose's violent vision of nationalist struggle meant he saw the INA as a military crusade to reclaim Delhi and drive out the British from India. However, in reality, the INA failed alongside the Japanese in the war, and captured soldiers were seen as traitors by the British. However, fighting in Burma, the RIN sailors came to see these men who were violently resisting colonial rule:

> It was in Rangoon that we came to know about [the] INA. Indians in Rangoon gave us INA literature and told us hundreds of Azad Hind Fauj episodes. [Ships operated] in the Malacca Straights and in the vicinity of Singapore we

> came to know all about its literature – ideals – patriotic deeds, and its camps and establishments. Every rating obtained some sort of matter containing INA Subhashe's [*sic*] photos and speeches, Azad [Freedom] Army's newspapers … Especially communication branch ratings had their kit bags full of such things including INA gramophone records. Lot[s] of ratings visited Bose['s] camps where Indonesian evacuees were held, and told countless fables of Indonesia ['s] revolt and British oppression. (Telegraphist Ahmed, Memo's of Witnesses Submitted to the Inquiry, 131).

Sailors smuggled pieces of literature on board their ships (see also Dutt 1971), but some INA soldiers were actually transported back to India on board RIN ships (Bose 1988). By being in these soldiers' company at sea – contained aboard the ship and coming into close contact with the INA men – for an extended period of time sailors began to be radicalised and to think about the potential for violent resistance to colonialism. This refashioning of solidarity by the sailors created a type of subaltern cosmopolitanism[5] amongst them. Rather than being 'indoctrinated' by the colonial naval authorities, the RIN sailors re-worked their loyalties towards other people in similar situations. Crucially, it was the extended mobility of their lives as sailors that allowed the sailors to come into contact with these groups. Whilst the rhythms and patterns of naval life on board ship were intended to maintain control, the physical movement of the ships themselves was instrumental in breaking down the ability of the colonial authorities to govern men. The very mobility of the RIN sailors as 'translocal subjects' (Conradson and Mckay 2007) of colonialism helped them to question and resist the disciplinary frameworks that naval and colonial life placed upon them. 'Passengering' (Ashmore 2013) INA soldiers across the Indian Ocean in the confines of the ship gave RIN sailors time to learn about nationalism and Bose's fight for Indian independence. However, a final level of (im)mobility which affected the men of the RIN lay in the times when the ships as material objects were at their least 'mobile': when they were in port.

As Dening (1992) argued, the ship in port is an even more liminal space. Whilst the notion of the fully self-contained world of the ship at sea is fanciful, being in port clearly allowed sailors to come into relatively more contact with social occurrences ashore. Unlike the eighteenth Century naval ship of Dening, by the 1940s advances in communication technology, together with the construction of subaltern cosmopolitanism detailed above, meant that the always tenuous binary of relations in and out of harbour is much less certain, as we have seen when RIN ships mutinied at sea when they heard of events ashore. This technological advancement meant that, despite the spatial and organisational structuring of the ship and RIN, the shoreline became blurred as relations between ship and shore become entwined.

With rapid demobilisation and poor administration occurring, the nature of life in the RIN in 1946 meant that many men lived in shore establishments, and even those sailors based on board ships had ready access to the mainland and the ideas circulating around India. During the Commission of Inquiry conducted after the Mutiny, officials were concerned about the types of literature that ratings had access to whilst on board. Copies of newspapers, nationalist speeches and ideas, and in some cases Communist literature were all reported as being found. Indeed, the ability of ratings to gain access to outside literature was seen as of key importance to the authorities. For instance, the Chief of Personnel in Delhi, Commander Jefford, when questioned about the 'gullibility' of the ratings to these subversive influences, argued that being poorly educated meant that sailors were more easily controlled as they would be unable to read newspapers (Delhi Proceedings, in Bombay Witness Statements, 71). However, this dismissive view of ratings ability to communicate amongst themselves was questioned by other officers in the inquiry, for instance, Lt. Krishnan, the Commanding Officer of HMIS *Shamsher*, one of the few ships not to mutiny, stated that:

> One cannot stop the men from taking an interest in politics, and a sympathetic understanding will pay more than repression – a rating will not write *Jai Hind*[6] on a bulkhead if it is explained to him that somebody else will have to rub it off. Far too much stress is made on subversive literature. If these pamphlets are allowed to flood the markets, how can you stop them from getting on board without a daily search of the men – a fatal thing to do. (Memo's of Witnesses, 261)

This more nuanced account of sailors' political knowledge, whilst still paternalistic, allows the sailors themselves a degree of agency, something which most authorities, thanks to naval doctrines stressing containment and order, together with the racialised stereotypes of docile or gullible colonial Indians, were unable to understand. In reality, as Krishnan states, newspapers were widely available to sailors whilst they were in port and in shore establishments. These papers communicated events such as nationalist speeches across India, but also the trials of INA men taking place in Delhi which attracted much hostility – in Karachi, sailors established a support fund to raise money to pay for INA soldiers' legal fees (Deshpande 1989). The political situation in Bombay, with widespread Communist Party activity (see, for instance Chandavarkar 1998) also meant that some ratings were aware of more left-wing political ideas which permeated through their ideas. BC Dutt, the sailor who was arrested for painting slogans on the walls of HMIS *Talwar* prior to the Mutiny, later wrote a memoir of his life in the Navy (Dutt 1971). In this, he details how he was politicised throughout his life in the Navy, including encountering INA troops in Burma and smuggling information back on board his ship. However, he also details how, once he was transferred to *Talwar*, he tried to recruit others to more radical causes, firstly in the canteen, but once sailors had shown an inclination to radical ideas, by taking them out drinking in Bombay (Dutt 1971, 78). Thus, both within the overcrowded and undisciplined naval base, but also within the urban spaces of Bombay, radical ideas were allowed to circulate. Although the numbers of radical sailors was small (Dutt claims there were only 20 working with him) the impact of these sailors, together with the more general circulation of nationalist ideas meant that political knowledge was widespread throughout the RIN.

This in turn meant that, particularly in Bombay, Navy officials were particularly concerned about the influence of popular politics 'ashore' upon the men of the RIN. This concern permeated through the service to the highest levels, as Admiral Godfrey, the Navy's Commanding Officer clearly showed in his testimony at the Commission of Inquiry:

> Since the war there has been a steady building up of quite a mass of quite fantastic and uncontrolled anti-British propaganda. The less responsible press preached this hymn of hate and in the case of certain journals in Bombay the Navy was singled out as a special target – largely due in one case to the

> connection between the staff of the paper and a disgruntled Indian officer. (Delhi Proceedings in Bombay Witness Statements, 93–94)

As a result, it is impossible to think about the RIN Mutiny, a naval event, without thinking through the connections that the RIN had to politics 'ashore'. These connections were made by sailors actively taking part in the circulation of people, objects and ideas like INA prisoners, newspapers and gramophone recordings of Subhas Chandra Bose. The ordered military ship and shore establishment-based spaces of the RIN had been subverted into both spaces of disorder and revolution (in the eyes of the colonial authorities) but spaces of political potential for the sailors, both by the organisational failures of the service, but also by political activity on board the ships and shore establishments of the navy. This reading then places the mobile lives of naval sailors into a world that does not end with the material entity of the ship, but extends them into wider assemblages of naval doctrine and colonial discipline.

Conclusion

The differential mobilities of the sailors of the RIN had crucial effects upon their political lives. On one level, disciplinary practices within the spaces of the Navy also created specific rhythms and experiences designed quite literally to 'indoctrinate' sailors' bodies into working effectively as a part of the assemblage that was the RIN. However, this disciplining process was contested by spatial and societal processes shaped by the movement across space that life in the RIN entailed – discovering what life was like in democratic nations, learning 'large ideas' from fellow sailors in the RN, and hearing radical stories promising a democratic future for India. As a result, the mobile relationality of 'the ship' must be read contingently as it is produced in wider societal interactions. Contra Dening (1992), rather than thinking that the ship in harbour is somehow 'more liminal' than one at sea, the relational and mobile world inhabited by the sailors of the RIN meant that social and political activities could permeate the ship wherever it was. Instead, as the men of HMIS *Shamsher* proved, mutiny could be avoided with a more considered relationship between officers and men, and especially when a reasonable proportion of officers were Indian.

The ships and shore establishments of the RIN, then, were spaces where the difference between coloniser and colonised played out. So, rather than being liminal only in certain places, such as the port, they were spaces where the disciplinary and 'civilising' notions of colonialism were lived and experienced by the ratings. For many sailors, and for much of the Second World War, these disciplinary practices were tolerated – BC Dutt, for example, speaks in his memoirs about how he enjoyed certain aspects of naval discipline and the camaraderie it inspired. However, when combined with a volatile political situation in mainland India and their exposure to the violent anti-colonialism of the INA, together with a generally poor standard of administration and organisation within the RIN, the violence of 1946 was increasingly likely.

Sailors' (im)mobile lives lay at the heart of how they felt and saw these processes of (in)justice at work at sea and ashore, and how they evolved their relationship to them. Thus, it was through experiencing (im)mobility aboard ships and in the RIN more broadly that sailors learnt to be 'political' by seeing the inequalities that lay at the heart of the colonial system – being unable to speak to RN sailors in the UK, transporting nationalist INA soldiers who were classed as traitors, and being racially abused as a part of shipboard discipline. Historical maritime spaces such as this expose the differential temporalities of mobility, and crucially how individuals experience different forms of mobility which are relationally constituted and productive of embodied and affective geographies of power. The RIN sailor was enmeshed in military rhythms and practices of discipline in the naval spaces which they inhabited, but the maritime nature of the mobilities experienced by these sailors, together with their growing knowledge of (and resistance to) colonial governance, all inherently shaped the political subjectivity that they went on to inhabit.

Crucially, these historical mobilities are more than simply 'slowing down' or lessening the intensity of (im)mobility compared to seemingly more urgent, speedier, contemporary mobilities. Instead, understanding the layers of mobility experienced over time allows mobility to be understood more extensively, with individuals' positionalities evolving as they moved through the spaces of the RIN over months if not years. Men did not arrive fully formed as sailors, radicals or mutineers, but instead negotiated their identity as they moved through the RIN and its spaces, being disciplined and resisting as they saw fit. As such, historical mobilities such as this have the potential to show not only how mobility has always been relationally embedded in wider societal, political, cultural and historical events, but crucially how these processual, long-term mobilities are productive of large-scale political events like the RIN Mutiny.

Notes

1. The term 'mutiny' is contested. For some sailors, this was a 'strike' for better working conditions, and a 'Central Strike Committee' was organised. For others, and particularly amongst the Indian Left's historiography of the event, the mutiny has been rationalised as a 'revolt' for Indian independence. See Davies (2013a) for more on this issue.
2. The term 'Ratings' refers to those sailors who are not 'commissioned officers'. It is therefore a blanket term for all ranks and branches of the naval service that are below officers, and includes 'non-commissioned officers' such as petty officers, who are the equivalents of sergeants or corporals in an army.
3. In naval nomenclature, shore establishments were known as 'ships', thus *Talwar* despite the designation HMIS was a naval shore establishment, and essentially was a base with a parade ground and barracks in downtown Bombay. Shore establishments varied according to their purpose as most were for specific branches of the RIN (hence *Talwar* trained signallers). Doctrine and naval life on ships and in shore establishments were very similar; however, sailors in shore establishments clearly had more opportunity to access the city compared to sailors on board a ship at sea.
4. The WRINs did not join the Mutiny in 1946 (Hastings 1988). Although there were women in service, most of the archival and published material consulted deals only with a distinctly masculine ideal of service. Therefore, I write this piece predominantly from a necessarily male-centric perspective. It would be useful for further research to address women's roles in colonial-armed forces in general as well as in the RIN.
5. Whilst it is difficult to claim that these men were 'subaltern' as many of the men were from relatively well-educated backgrounds, and by working for the colonial state, they had access to the authorities which more 'subaltern' groups did not, their ability to negotiate their position in relation to other groups facing similar struggles against the colonial order allows them to be thought of as constructing a form of subaltern cosmopolitanism.
6. Jai Hind, meaning 'Victory to India' was a common nationalist slogan.

References

Unpublished Sources

ADM 1/19411. *Dominions, Colonies, Protectorates and Mandated Territories Mutiny in the Royal Indian Navy 18–26 Feb: Reports from C in C East Indies Station and Flag Officer*. London: Royal Indian Navy National Archives.
Bombay Witness Statements from the Commission of Inquiry – Transcripts RIN Mutiny Papers – Serial No. 18 National Archives of India. New Delhi.
Government of India. 1945. *Creeds and Customs of the Royal Indian Navy British Library*. London
Folder Containing Memo's of Witnesses Submitted to the Commission of Inquiry RIN Mutiny Papers – Serial No. 13 National Archives of India. New Delhi.
Karachi Statements RIN Mutiny Papers – Serial No. 14 National Archives of India. New Delhi.
Newspaper Clippings Bearing on the Mutiny RIN Mutiny Papers – Serial No. 8 National Archives of India. New Delhi.
Report of the Commission of Inquiry (3 Volumes) RIN Mutiny Papers – Serial No. 6 National Archives of India. New Delhi.

Published Sources

Adey, P. 2006. "If Mobility is Everything Then It is Nothing: Towards a Relational Politics of (Im) Mobilities." *Mobilities* 1 (1): 75–94.
Ashmore, P. 2013. "Slowing down Mobilities: Passengering on an Inter-war Ocean Liner." *Mobilities* 8 (4): 595–611.
Bearman, P. S. 1991. "Desertion as Localism: Army Unit Solidarity and Group Norms in the US Civil War." *Social Forces* 70 (2): 321–342.
Bose, B. 1988. *RIN Mutiny 1946*. New Delhi: Northern Book Centre.
Chandavarkar, R. 1998. *Imperial Power and Popular Politics: Class Resistance and the State in India, c. 1850–1950*. Cambridge: Cambridge University Press.
Chandra, B., M. Mukherjee, A. Mukherjee, S. Mahajan, and K. N. Panikkar. 1989. *India's Struggle for Independence*. New Delhi: Penguin.
Conradson, D., and D. Mckay. 2007. "Translocal Subjectivities: Mobility, Connection, Emotion." *Mobilities* 2 (2): 167–174.
Corona, V. P., and F. C. Godart. 2010. "Network-domains in Combat and Fashion Organizations." *Organization* 17 (2): 283–304.
Cresswell, T. 2010. "Towards a Politics of Mobility." *Environment and Planning D: Society and Space* 28 (1): 17–31.
Das, D. K. 1994. *Revisiting Talwar: A Study in the Royal Naval Uprising of February 1946*. New Delhi: Ajanta Publications.
Davies, A. D. 2013a. "Identity and the Assemblages of Protest: The Spatial Politics of the Royal Indian Navy Mutiny, 1946." *Geoforum* 48: 24–32.
Davies, A. D. 2013b. "From 'Landsman' to 'Seaman'? Colonial Discipline, Organisation and Resistance in the Royal Indian Navy, 1946." *Social & Cultural Geography* 14 (8): 868–887.
Deleuze, G., and F. Guattari. 1987. *A Thousand Plateaus*. London: Continuum.
Dening, G. 1992. *Mr Bligh's Bad Language: Power, Passion and Theatre on the Bounty*. Cambridge: Cambridge University Press.
Deshpande, A. 1989. "Sailors and the Crowd: Popular Protest in Karachi, 1946." *Indian Economic & Social History Review* 26 (1): 1–28.
Driver, F. 2006. "Shipwreck and Salvage in the Tropics: The Case of HMS Thetis, 1830–1854." *Journal of Historical Geography* 32 (3): 539–562.
Dutt, B. C. 1971. *Mutiny of the Innocents*. Bombay: Sindhu Publications.
Featherstone, D. J. 2009. "Counter-insurgency, Subalternity and Spatial Relations: Interrogating Court-martial Narratives of the Nore Mutiny of 1797." *South African Historical Journal* 61 (4): 766–787.
Foucault, M. 1986. "Of Other Spaces." *Diacritics* 16 (1): 22–27.
Foucault, M. 2007. *Security, Territory, Population: Lectures at the College De France, 1977–78*. Basingstoke: Palgrave Macmillan.
Gogia, N. 2006. "Unpacking Corporeal Mobilities: The Global Voyages of Labour and Leisure." *Environment and Planning A* 38 (2): 359–375.

Gourgey, P. S. 1996. *The Indian Naval Revolt of 1946*. London: Sangam Books.
Hastings, D. 1988. *The Royal Indian Navy, 1612–1950*. London: McFarland & Company.
Hasty, W. 2011. "Piracy and the Production of Knowledge in the Travels of William Dampier, c. 1679–1688." *Journal of Historical Geography* 37 (1): 40–54.
Hasty, W., and K. Peters. 2012. "The Ship in Geography and the Geographies of Ships." *Geography Compass* 6 (11): 660–676.
Hyslop, J. 2009. "Guns, Drugs and Revolutionary Propaganda: Indian Sailors and Smuggling in the 1920s." *South African Historical Journal* 61 (4): 838–846.
Land, I. 2006. "Sinful Propensities: Piracy, Sodomy and Empire in the Rhetoric of Naval Reform, 1770–1870." In *Discipline and the Other Body: Correction, Corporeality, Colonialism*, edited by S. Pierce and A. Rao, 90–114. London: Duke University Press.
Law, J. 1986. "On the Methods of Long Distance Control: Vessels, Navigation, and the Portuguese Route to India." In *Power, Action and Belief: A New Sociology of Knowledge?* edited by J. Law, 234–263. Henley: Routledge.
Law, J., and A. Mol. 2001. "Situating Technoscience: An Inquiry into Spatialities." *Environment and Planning D: Society and Space* 19 (5): 609–621.
Madsen, C. 2001. "British Officers and Striking Sailors: Mutiny in the Royal Indian Navy, February 1946." *American Neptune* 61 (3): 299–315.
McKee, C. 2002. *Sober Men and True: Sailor Lives in the Royal Navy 1900–1945*. London: Harvard University Press.
Nordstrom, C. 2007. *Global Outlaws: Crime, Money and Power in the Contemporary World*. Berkeley: University of California Press.
Ogborn, M. 2002. "Writing Travels: Power, Knowledge and Ritual on the English East India Company's Early Voyages." *Transactions of the Institute of British Geographers* 27 (2): 155–171.
Ogborn, M. 2008. *Global Lives: Britain and the World 1550–1800*. Cambridge: Cambridge University Press.
Ong, C.-E., C. Minca, and M. Felder, Forthcoming. "Disciplined Mobility and the Emotional Subject in Royal Dutch Lloyd's Early Twentieth Century Passenger Shipping Network." *Antipode*.
Ryan, J. 2006. "'Our Home on the Ocean': Lady Brassey and the Voyages of the Sunbeam, 1878–86." *Journal of Historical Geography* 32 (3): 579–604.
Spence, D. O. 2014. "Imperial Transition, Indianisation and Race: Developing National Navies in the Subcontinent, 1947–64." *South Asia: Journal of South Asian Studies* 37 (2): 1–16.
Wald, E. 2012. "Health, Discipline and Appropriate Behaviour: The Body of the Soldier and the Space of the Cantonment." *Modern Asian Studies* 46 (4): 815–856.

Unraveling the Politics of Super-rich Mobility: A Study of Crew and Guest on Board Luxury Yachts

EMMA SPENCE

Cardiff School of Planning and Geography, Cardiff University, Cardiff, Wales, UK

ABSTRACT *In this paper, I introduce the superyacht as a unique vessel, as a home and workplace to professional crew and holiday space for its super-rich passengers. Drawing upon the notions of* motive, rhythm, *and* friction *from Cresswell's mobility constellation, this paper illustrates how the politics of super-rich mobility are performed by crew and guests on board. Rather than preserve the perception that super-rich individuals are hyper-mobile I in turn suggest that in the case of the superyacht the desire and to perform their mobility status when on board ultimately circumscribes or restricts super-rich mobility. Using data from an in-depth ethnographic study on board various superyachts, I suggest ways in which the mobility of the yacht can inform both super-rich and shipped mobilities as both fields continue to grow.*

Geographies of Ships

Container ships, passenger ferries and cruise liners are tied to schedules and predetermined and meticulously planned itineraries that guide the vessel, its cargo, crew and passengers from port to port. Yet ships are not empty vessels that traverse watery soulless highways, rather ships with their crew, passengers, human and/or non-human cargo, are lived and experienced places. In recent years, growing interest in geographies of the sea has challenged this terracentrism of geography and the social sciences (see Peters 2010; Blum 2013; Steinberg 2013). However, increasing concern with both the contemporary and the historical global movement of goods and people has somehow failed to simultaneously emphasise the importance of vessels – with the ship remaining an elusive, and largely forgotten space in social studies (Hasty and Peters 2012, 660). The relative lack of contemporary investigation in to the ship (with the noted exception of Hasty and Peters 2012) neglects the lived experiences of those on board as a significant area of study. As well as a tool to exploring wider themes of commerce, globalisation, and time-space compression, the study of ships and their various and complex mobilities presents multiple avenues of enquiry for scholars, particularly in terms of social-spatial relations. Factors such as strict hierarchies of crew, watchkeeping and shift work directly impacts upon how people on board not only relate to each other, but also to the ship itself, the sea surrounding them and the shore that they have left behind, or indeed are heading towards. In this paper, I introduce another, very different, type of ship – the luxury superyacht. Instead of linear and rigid itineraries familiar to commercial shipping (see Martins 2013 for an excellent review of the mobility of shipping containers), the mobility of the yacht is largely determined by the whimsical demands of its super-rich passengers (hereafter guests). In addition, the ratio of crewmembers to guests on board, a yacht is close to 1:1 (depending on length and size of the vessel) which enables a more personal and attentive service to the guests and which thus creates different social-spatial interactions compared to large commercial passenger vessels such as cruise ships. Exploring the interactions between crew and guests on board, the luxury yacht highlights how each group experience mobile practices differently. Using the example of the superyacht presents numerous avenues of geographical enquiry that go some way to reinforce the significance of researching the mobility of ships which will be further explored throughout this paper. For instance in considering the cultural practices of the super-rich helps to identify and account for the group's global mobility can help to stimulate further super-rich and hypermobile debates within mobilities studies (see Birtchnell and Caletrio 2013; Hay 2013 for super-rich mobilities). Following Cresswell, the mobility constellation disentangles these complex social-spatial politics, with a focus on six aspects of mobility: 'motive', 'speed', 'rhythm', 'route', 'experience' and 'friction' (2010, 17). In this paper, I use specifically the notions 'motive', 'rhythm' and 'friction' in order to disentangle the politics of super-rich mobility between guests and crew on board these ships. Thinking through mobility within these divisions allows mobility scholars to dissect unequal power relations that ultimately shape patterns of physical movement (Vannini 2011, 481). In this paper, I argue that the motives and rhythms of mobility experienced by crew and guests on board is continuously negotiated performance between the two groups. Super-rich mobility is performed firstly by super-rich guests to their peers, and secondly by crew to create a spectacle of super-rich mobilities for their guests. Conversely, it is this desire to perform that circumscribes super-rich mobility when on board as their activities, encounters and practices are directed at specific venues and port towns in a limited geographic area.

A Note on Methods

Differentiated mobility, between a mass of human and non-human actors, interacting in innumerable ways is difficult to articulate and even more complex to study empirically. This difficulty is exemplified in the study of three hyper-mobile subjects; in this case, the luxury yacht, the super-rich guests and the yacht crew. To overcome these issues, I have collated a variety of data sources from ethnography, interviews, and social media and internet blogs in order to compile a series of vignettes. Utilising my experience working in the superyacht industry as crew over the past several years, I have opted for a largely ethnographic approach to studying the cultural geographies of super-rich mobility. My research project, from which this paper derives, looks at the networks that support and facilitate elite hyper-mobility, rather than the super-rich individuals directly, which is reflective of my position and experience as an early career researcher. The network of actors who facilitate super-rich mobility, such as crew explored in this paper, provides valuable and significant insight in to the cultural practices and processes of mobility enjoyed by the global elite. To compile a more holistic review of the politics of super-rich mobility on board yachts, I acknowledge that further investigation and interaction of super-rich individuals themselves is needed in future study. My ethnographic observations have been recorded in a montage of written field diary entries, scratch notes, film and photographs. Names of respondents have been changed to protect anonymity. My observations have been supplemented with interview data conducted with superyacht industry workers (yacht crew, shore-side provisioners, crew agents, air crew and yacht brokers).

The Politics of Mobility on Board

In mobilities studies, *mobility* refers to the ontology of movement – one that presents a critical approach towards the relations between acts of movements, the meanings invoked, the politics involved, and the social and physical implications of those relations as a result of acts of movement. *Motility,* on the other hand, expresses the ability to move. That is the access to movement, the motives and frictions to movements, and an individual's ability or inclination to enact this motility. As Kaufmann explains his reading of motility is that of 'a propensity to be mobile … which is likely to vary in intensity from one person to another' (Kaufmann 2002, 39). *Movement* thus refers to the actual physical act of moving. Whilst mobilities studies are generally concerned with representing flows and interactions, mobility is inherently political (Baerenholdt 2013, 20). The politics of mobility refer to the ways in which mobilities shape – and are shaped by – social relations (Cresswell 2010, 21). The politics of mobility is thus central to the idea that mobility is differentiated – meaning 'different things, to different people, in differing social circumstances' (Adey 2006, 83). As this paper will go on to outline, the various *motives* of travel for passengers and crew, the *rhythms* and routines on board, and the *frictions* to mobility encountered on board each of various vessels differ greatly from the other. The politics of mobile practices are illustrated in such cases where different groups of actors experience the same mobile practices in opposing ways (Cresswell 2010, 21). On board, the yacht guests dictate their daily mobilities including what ports they would like to berth in or what anchorages they would like to visit. These demands are met by the crew who collectively work 24 h a day to ensure that the guests' needs are catered for.

Scholars pioneering the contemporary and emerging field of super-rich mobilities have recently called for greater empirical depth in our understanding of the nature of super-rich mobility and the 'kinds of services provided by the various companies servicing the mobility needs of the super-rich' (Beaverstock and Faulconbridge 2013, 58). In response, this paper is written from the perspective of yacht crew, which enables valuable insight in to the exclusive world of the geographies of super-rich mobility.

Ranging from anything between 25 and over 100 m in length, multi-million pound luxury yachts are owned and used by the global super-rich and operated by highly trained, professional yacht crew. The yacht is a complex space of work, home and leisure space as crew carves a sense of home and belonging in the vessel that predominantly serves as their workplace. Workplace geographies explore the social-spatial power relations within the workplace. Sayer and Walker (1992) explain how the 'basic organising principle of the workforce is containment within a limited area' (cited in Crang 1994). The restricted physical area of the boat, exemplified when at sea, serves to border and confine crew more explicitly than those who are employed in terrestrial workspaces. Exploring workplace emotions, Fineman (2003) explains that upon entering the workplace 'we bring our loves, hates, anxieties, envies, excitement, disappointments and pride' (Fineman 2003, 1). It is easy to relate to these personal workplace emotions, but on board, the 'loves', 'hates' and 'anxieties' are not detached from the crewmember's work or home experiences. On board emotions are constructed within the same small group of people in the same limited environment. There is little escape from the confines of the boat, which makes the working and living experiences intense. The distinction between workplace and homeplace on board is blurred therefore the crew have to control their emotional traces on the boat in order to remain professional in what is ultimately, a place of work. For example, if the crew are able to control or submerge their true emotions (perhaps stress, tiredness and irritability), it is then necessary for them to *perform* positive actions (smiling, politeness and feigning interest) to ensure the emotional maintenance of the boat for the benefit of the rest of the crew and the paying guests. As crew are unable to step outside of take a break away from the workplace to relive tensions as we may do ashore, I asked several respondents what they felt was different about working on board a ship compared to traditional terrestrial spaces such as offices. Debbie, an interviewee, had been employed in a city office before working on board as a yacht stewardess and she responded:

> You live and work there [the yacht] for starters. Then you wake up somewhere different most days … To wake up and not know where you are or where you are going becomes trying after a while.
>
> Debbie, stewardess, interview respondent

Continuously mobilised on board a yacht with no say in your own pace, direction or destination of mobility can become tiresome as Debbie felt. Her response illustrates the lack of power or influence that she has on the mobility of the yacht and therefore of her own mobility. Despite the tensions involved with continuously moving between ports and anchorages, and being completely at whim to the demands of the guests, their employment on board luxury yachts can spurn further personal mobility for crew when opportunities to go ashore arise. As Cresswell suggests: 'one person's speed is another person's slowness. Some move in such a way that others get fixed in place' (Cresswell 2010, 21). However in the case of the luxury yacht, rather than impeding the mobility of those around them, the hyper-mobile super-rich can actually enhance the personal mobility of the yacht crew. With no need to pay for food, rent, amenities, even toiletries when on board, and with the regular tips and bonuses, yacht crew are able to save substantial amounts of money when working on board this enables crew to travel extensively when permitted holiday on board. This desire and ability to travel is fuelled by diverse nationalities employed in the industry, presenting new opportunities and inspiration for travel when off the boat. As the following response from Joe a yacht engineer illustrates:

> Last month we pulled in to port at 9 o'clock at night, having dropped off guests that afternoon. At 9:15 the Captain confirmed that we could have the rest of the weekend off. Quick as a flash me and another crewmember jumped on the web and booked a weekend trip to Rome. It was all booked by 10 pm, to fly out at 6 am the following morning. Just enough time in between for a night out in the port bar. It's about making the most of every moment. Joe, yacht engineer, interview respondent

In this example, having been given the all clear from the Captain, Joe seizes the opportunity to control his own mobility when the guests or owner of the vessel are not present to demand his time and service. Rather than relaxing for an evening after what would have been a work-intensive week, Joe preferred to fill the time before catching his early morning flight socialising with friends away from the boat whilst he could. Although clearance for travel was required from the Captain and that the crew are still under the control and direction of somebody else, this power is more easily negotiated between crew and Captain as, whilst still the immediate boss, the Captain also lives and works on board with the rest of the crew.

As the guests dictate the ship's movements in terms of preferred locations to anchor or berth this implies that the crew are passive subjects in the mobility of the super-rich. However, this is a rather simplistic account of the politics at play on board as the crew themselves retain their own agency by opting to be on board in the first place. Their motives for facilitating super-rich mobility (such as high wages, low-living costs, travel experiences and maritime careers) contrast with but are as important as the motives of the guests (holiday, leisure, events and business meetings).

Untangling the Politics of Super-rich Mobility: Motive

Motive explores the original motivations of incentives that encourage or enable people to move in the first place (Cresswell 2010, 22). In this section, I argue that the primary motivations for super-rich to use yachts is to be able to perform their mobility status. As performing mobility is 'quintessential to personal and collective identity' (Vannini 2011, 480), owning or chartering a luxury yacht is a way for elites to showcase or to reaffirm their super-rich membership. The failure to own or to regularly charter a yacht could be viewed as 'lack of competence and illegitimacy' as a member of the elite group (Beaverstock and Faulconbridge 2013, 52). As an example, when opting to go ashore guests prefer to go to prominent ports, with reputable bars and restaurants and to associate themselves with other super-rich individuals. As such, the privacy and exclusivity offered by luxury yachts is somewhat juxtaposed by the need for to perform their mobility status (Beaverstock and Faulconbridge 2013, 55). In contrast to the super-rich the motivations for crew to seek employment on board are typically to earn a good wage with little or no living costs, and the opportunity to travel to otherwise inaccessible places. With two very different motives for being on board a yacht, the two groups inevitably have different experiences and different personal mobilities on board. It is the job of the crew to construct and to help perform displays of super-rich mobility for their guests. The following vignette compiled from a montage of ethnography, and interviews conducted on board and ashore outlines the 'behind the scenes' performance preparations that precede the arrival of chartering yacht guests.

We're told that the guests' plane lands at 11:15 at Nice Airport. That means we can expect them on the dock in Cannes at 12:30 at the earliest. We need to change in to our smart 'number ones' by 11:30 (black shirt, epaulettes, black trousers, black shoes for interior, white shirt, epaulettes, black trousers, black shoes for deck crew). Before I change, the main deck stairs need to be vacuumed. Shani only did them yesterday and not a soul has walked on them since, but just incase a stray hair or piece of fluff has found its way on to the cream carpet, it's best to vacuum again. Shani busies herself with setting up a display of fruity ice tea and fresh cherries in the upstairs lounge. Sophie neatens the flower displays on the outside tables. Amy makes last minute amendments to the accounts before she does a last sweep of the interior to make sure everything is in place and ready.

11.15 and the guests land at Nice airport. The Captain awaits them inside the private aviation terminal. The drivers of two blacked-out Mercedes vans wait to transport the guests to the yacht. The drivers are dressed smartly in black suits, white shirts, and black ties. The Captain meets the guests in his four-stripped epaulettes, crisp white shirt, and shiny black shoes. He shakes their hands firmly, smiles broadly, and oozes an air of importance. Pleasantries dealt with, the guests stride toward the Mercedes vans and the Captain strides faster, ready to pull open the passenger doors. The drivers scramble to swiftly load all of their bags into the second van. As the vans pull away and merge on to the A8 the Captain is in full flow: 'My crew are excellent, only the best for our guests … what a good week to charter, the weather is going to be fantastic'. The Captain slips out his Blackberry and texts back to the yacht: '7 guests, 16 pieces of luggage, eta 30 minutes'.

30 minutes. I've already changed but I'm lingering in my cabin. What is left to do? I mentally run through the list … lights, music, champagne! I place 6 champagne glasses on a silver tray and leave it in on the bar. I open the bottle of vintage *Dom Pérignon* and place a raspberry in to each glass. It's okay to open the champagne now but I cannot pour it until the guests are eta 3 minutes, as it needs to be as bubbly as possible. Music is always a tricky one for new guests. You want to set an atmosphere, but not knowing what the new guests like make it hard to choose. Something upbeat but that will fade in to the background. A moment later Jack Johnson's voice is booming out from every room and deckspace on the boat. He's then swiftly replaced by a Café del Mare compilation. Not all the crew like Jack Johnson, so Café del Mare is a good, neutral compromise. After all, we're the ones having to listen to it all day.

Our radios crackle in unison as the Captain announces his imminent arrival: 'eta 3 minutes'. The deck crew rush to line the dock, arms folded neatly in front of them, ready to shake hands and take care of the luggage. I pour the 6 glasses of champagne. There are 7 guests but there's bound to be a nanny, a mother, or a PA who doesn't drink, in which case they'll help themselves to a bottle from the water basket. Lights are on, doors are open, music is playing, and 6 freshly filled champagne glasses stand tall on the polished silver tray. The interior crew splits in to two lines, one either side of the doors that the guests will use to enter the interior. On the adjacent sea wall there's a man,

> mid fifties, wearing just a small pair of shorts, bare belly spilling over the top, who stops and examines the sudden rearrangement of crew. A young mother and her two small children also pause on the wall, hands shielding their eyes from the glare of the sun as they squint up the dock in anticipation of our new arrivals. Hands crossed neatly together, smiles on all our faces, we await our guests. The chef, dressed in his immaculate chef whites, abandons his stove and races down the port side to make it to the aft deck, clasping his hands in front of him, smile on face, just in time. Two black Mercedes vans slow then stop. All stood smiling, the drivers rush to pull open the van doors. Out step our guests for the week. The Russian elite.

The performance of greeting the guests, with the crew arranged in formation, dressed in their best uniform, with hair neat and faces shaven, serve to dramatise the guests' arrival to the yacht. The exclusive arrival at the private terminal, the Captain's enthusiastic spiel, the 'covert' radio communications, and even the small crowd gathered on the sea wall, all combine to make the guests arrival a performance. The finishing touches, such as the exterior flower displays, the bowl of iced cherries and champagne on arrival may not seem like extravagant gestures in themselves, but combined they make the super-rich guests feel important and fuels their motivations to board a yacht. Preparing the additional touches and changing in to a smarter uniform and making the effort to polish our shoes are integral to ensure the best first impressions are made to the new paying guests. It is the role of the crew to help the super-rich perform their mobility by means of the yacht.

In the Cote d'Azur region of the South of France, when opting to go ashore yacht guests are motivated to go to well-known ports, with exclusive bars and restaurants and more super-rich individuals, whereby they can see and be seen, with the yacht playing an integral part to the performance of mobility. Prominent and exclusive ports such as Monaco, Cannes and St Tropez along the Cote d'Azur attract higher fees for berths and are always in high demand from the yachting super-rich. The exclusivity of Monaco for example, with its members-only casino, restaurants, hotels and bars ensures that the town caters for and continues to attract a stream of super-rich and their yachts. Despite retaining their allure to the super-rich, regular exposure to the exclusive shore spaces such as Monaco, Cannes or St Tropez can grow less impressive overtime for the crew. As the following extract shows, whilst guests enjoy the spectacle of their arrival via yacht in to prominent ports, the crew experienced it somewhat differently.

> The big boat harbour in San Tropez is right in the middle of town and we berthed stern-to with a large audience of afternoon strollers. I was totally mortified and embarrassed to be a part of that huge ostentatious spectacle. I was used to the boat and didn't see it as being anything special, or even nice, and was momentarily stunned by the enormous crowd of onlookers it attracted. The more courageous members of the crowd were asking, 'Who's he? Who's he?' as if [the owner] was some kind of holy man or king. (Light 1999, 131)

In the typical working day for yacht crew, there are multitude actions and rituals to be performed, from the presentation of coffee or cocktails, to the changing of uniforms and flags to correspond with the rising or falling sun. Despite being a technical and at times difficult process, docking the yacht in a prominent port such as St Tropez, as described by Light, is also another performance by the crew. Whilst docking the crew are under pressure to protect the yacht as it backs in to a usually tight spot between two other multi-million pound yachts. They fall under the scrutiny of their colleagues from other yachts in port, and they are also under pressure from the watchful eyes of their bosses to make the perfect entry in to port, not to mention to the crowd of 'yacht-spotters' gathered on the dock. In Light's account the guests enjoy entering the prominent port and being a part of or even the cause of the spectacle. With so many people gathered on the dock willing to give their time and interest in trying to find out who is on board is a strong motivation for guests to board in the first place. Whereas most crew would be used to this display of docking as just another role to perform, in Light's case she as embarrassed by the spectacle; her motivations of being on board simply to earn a wage. The motivations for mobility in this instance are much different between guest and crewmember. With the guests looking to relax and enjoy their time on board and the crew primed to meet their every demand, the *rhythms* between the two groups on board are played out in a variety of ways.

Untangling the Politics of Super-rich Mobility: Rhythm

The spaces in which we inhabit and mobilise through are composed from myriad 'rhythms, temporalities, pacings and measures' (see also May and Thrift 2001; Edensor and Holloway 2008, 483). The rhythm of a particular spatiality can refer to the composition of repeated movements and rest within a particular space, and the embodied, sensual and affective qualities thus produced (Edensor and Holloway 2008, 483; Cresswell 2010, 23). Following Lefebvre's concept of rhythmanalysis (2004), I explore in this section the competing and complimentary rhythms simultaneously produced and experienced by yacht crew and guests. Given the yacht's confined spatiality, confounded by being at sea, and with the intense social interactions derived from a strict hierarchal ordering, the rhythms and pacings experienced and negotiated on board result in significant micro social-spatial politics. The internal rhythms, routines and practices on board the yacht are complex as, with guests on board, the crew are at their most busy, whilst the guests are typically at their most restful. In the following example, Annie a yacht stewardess, illustrates the differences in rhythm between guests and crew.

> When people ask me what I do for a job & I answer I'm a yachtie: they say ooh lucky you – you're living the dream! Yesterday I started work @ 8 am & it's now 5 am & I have just finished for the day after cleaning up vomit & other things from the 21 year olds onboard … It's not always about living the dream …
>
> Annie, interview respondent

In this example as the young guests on board party through the night, Annie continues to work, with their late night resulting in her exceptionally long shift. The way both the guests and the stewardess experienced this night differed in sights, smells, emotions, motivations, energy and interactions. However, despite the rhythm being completely different in order for the guests to continue their enjoyment of the yacht their rhythms were not out of sync but complimentary. Annie's fast-paced work-orientated rhythms enabled the guests' relaxed, intoxicated experience of the night on board the yacht.

The following extract from the author's field diary illustrates how the social-spatial rhythms on board can differ between crew and guests but also between crew as the mobilisations requested by the guests further attributes to the lack of power or control experienced in my own ability to influence personal rhythms when on board.

> I'm startled awake by a series of loud clanging noises. Metal on metal. My cabin is in the forward bow, so is in close proximity to the anchor locker. The loud clanging signals a new destination, a new anchorage, and more importantly signifies more noise to come. As the anchor is released from the locker the heavy chain spills from the bow, plowing the anchor in to the seabed. As the chain is drawn out to allow for the yacht to move around the anchor, the metal upon metal continues to shake through my cabin. The rush of chain reverberates through the steel hull of the boat, shaking my bunk to ensure that I'm by now fully awake. I pull myself from my bed and clamber on to the top bunk to look out of the porthole. As I anticipated we've anchored outside Nikki Beach. I hope we stay here for the rest of the day as I'm aware that the noise and vibrations of the anchor being lifted back in to the locker are even more prolonged and I'd really like to get back to sleep before the start of my next shift. Author's field diary entry

This extract highlights the differing spatial-temporal rhythms experienced between the crew, in this example the author is sleeping in the day following a night shift, whereas the remaining crew are working the day shift, which results in a conflict of patterns and rhythms on board. The noise of the anchor chain wakes me in my bunk, but so does the chatter and cupboard banging that occur during the lunch hour. As a result of shift work and staggered break times, the rhythms between crew can be greatly skewed too as some are mid-shift whilst others, like myself in the extract above, are asleep following a night shift. The difference in relaxed guest rhythms and rushed working crew rhythms are identified numerous times throughout the day. When guests sit down to lunch on the aft deck for example, the interior crew are at their most busy time of the day as with all guests outside and unlikely to leave the table once eating their cabins are available for cleaning. It is not unusual for a stewardess to be frantically drying the marble shower in a guest cabin during lunch to be called away by radio between courses to assist with serving food platters or to replenish wine glasses, only to rush back downstairs to finish the bathroom before dessert. During the evening dinner, there are more courses to serve, sometimes more guests to cater for, and several wine choices to keep abreast of. As the guests are guaranteed to be at the table for at least an hour there are showers to clean, carpets to vacuum and beds to 'turn down' (removal of the day cover, fold down the top sheet, iron out any creases, and lay an evening gift or chocolate on the pillow). As this section has outlined, the differences in how the two groups experience the mobility of the luxury yacht is reflected in their different rhythms on board.

Whether tied to the dock, anchored off shore or in transit to the next port town, the yacht and its crew are always in motion. However, there are times when the yacht is required to be entirely immobile.

Untangling Politics of Super-rich Mobility: Friction/Immobility

Mobilities studies challenge the long-standing relatively 'a-mobile' nature of social science thinking (Hannam, Sheller, and Urry 2006, 5). It seems converse therefore to talk of the need to explore the static qualities of mobility. However, immobility is central to our understanding of movement, as pauses of people and goods can be just as significant as their impact whilst in motion (Cidell 2012, 233). Immobility, or friction as a key element of the mobility constellation, refers to the 'barriers that stop, impede, or inconvenience our movement' (Vannini 2011, 472). Whether the function is to transport commodities, passengers or the world's wealthiest individuals, the very purpose of any vessel is to mobilise across a body of water. External relations such as weather systems, tides and lunar cycles have a profound affect on how the ship is experienced and mobilised (see Lambert, Martins, and Ogborn 2006 for materialities; Jones 2010 for lunar rhythms). Seasickness for example is a result of adverse weather conditions (see Peters 2012; Spence 2014 for a discussion on seasickness). Removing the ship from the sea however the most extreme impediment to the function of the vessel and how it experienced by those onboard. As Hasty and Peters ask: 'Can the ship ever be immobilised when, even at anchor, it is subject to the sway produced by wider relationalities with the sea and wind?' (Hasty and Peters 2012, 670). In the case of luxury yachts, the answer is yes; the ship can be and is often immobilised. At least once every two years, a typical yacht needs to be removed from the sea and placed in dry dock in order to be surveyed and to undertake essential maintenance work. Intrigued about how the immobility of the yacht impacts upon its crew, I visited a dry dock in Marseille in the South of France. Arriving at the dry dock and seeing the luxury yacht out of the water for the first time was impressive as you got a real sense of the size and grandeur of the vessel. The following vignette compiled through ethnographic observations and input from the crew on board, gives a sense of the impact of the immobility of the yacht for those who continue to live and work onboard:

> The boat doesn't feel like it should when it is out of the water. This may sound obvious, as it is in fact stationary – the opposite of what was intended by its creation. But it's more than this; the stillness of the boat seems to somehow stifle the experience of the boat for us crew. Materially, the boat is dirty, inside and out from the dust of the shipyard, which takes away from the impressiveness of the usually pristine vessel. It looks somehow fragile stood up on wooden stilts in a giant hole in the floor and having to climb down several sets of wobbly steel steeps to get on or off the boat seems wrong. It is our job to keep the boat looking, well, ship-shape, and not being able to do so is disheartening. It isn't just the physical detachment from the sea that seems to subdue the experience of the yacht. The absence of the gentle lapping against the hull as you sleep is magnified in its silence. We itch to get back in the water, to get the boat back to what and how it should be. Vignette compiled from interview with crew and author's ethnographic observations.

The yacht still functioned for the crew as it was still a place of work and still a place to call home, yet the immobility of the vessel really impacted on the feel of the yacht and after several dry weeks it some became obvious the importance of being at sea. When asked to reflect on the absence of the sea Graham, an officer on board, gave an insight in to how crew relate to the ship out of its usual marine surrounds, he says:

> The very nature of the sea allows us to be professional – to have a profession. There are laws to be adhered to, there are rules to follow, there are procedures, and safety concerns that are fundamental to operating a boat … It's the same as being a pilot I guess. Without the air to fly in, there's not much use for the plane.
>
> Graham, Officer, Interview respondent

Out of the water the yacht is clearly immobilised, which as the crew vignette illustrates has a profound affect of how the ship is experienced by those on board. The lack of water and therefore the lack of mobility transforms the feel of the ship. Furthermore in the case of the yacht even when the vessel is at sea, its mobility can be somewhat restricted. Being at anchor requires constant monitoring of the yachts position, with a screen in the bridge dedicated to tracing the pivot of the yacht around the anchor. The movement of the yacht is tracked on the screen, with any deviations of the semi-circle indicating that the anchor may have dragged and thus the yacht would be no longer secure. At anchor then the yacht continues to be mobile as it moves with the natural currents and winds, yet is fixed to a strict circumference thus circumscribing the ship's mobility. Furthermore, there is certain monotony to super-rich mobility on board a yacht goes some way to restrict or circumscribe the ship's mobility – much like the role of the anchor. Through the summer in the Mediterranean the majority of yachts cruise between the ports of St Tropez and Monaco, and often as far as Porto Fino on the Italian Riviera. This route, stopping at ports such as St Tropez, Cannes, Antibes, Monaco and numerous anchorages in between is affectionately known as 'the milk run' by crew in the industry. The routes and destinations of guests are easily predicted in spite of nationality, age or group type (friends, businessmen and families) of guests. For instance, a midsummer Saturday night in the Mediterranean means either Cannes or St Tropez depending on which exclusive nightclub has the better DJ that evening, followed by a Sunday morning dash to anchor at Nikki Beach (west of St Tropez) for watersports and deck parties. With nightclub opening and closing nights, exclusive events such as the Cannes Film Festival or the Monaco Grand prix, the summer calendar shapes the motivations and mobilities of super-rich individuals who have chartered or who own a yacht. In essence the port towns motivate super-rich individuals to use a yacht, as entrance by yacht enables them to perform their membership of the elite group, and to perform their holiday rhythms and rituals to their peers. In essence these port towns and anchorages where fellow super-rich yacht-goers travel to act as anchors to the ship's mobility. This circumscribed mobility is not only unique to the South of France, as particular events or dates influencing the location of superyachts as desired by their guests. St Barths in the Caribbean welcomes hundreds of the world's wealthiest for New Years Eve fireworks, whilst Richard Branson's Necker Island is rarely without a crowd of yachts during the European winter. Therefore, despite having the potential for hyper, friction-free mobility when onboard a yacht the guests tend to limit the mobilisations within a small geographical area. Although far from the reliability of a cargo ship or cruise liner itinerary, the use of yachts in this area does not reflect the potential for mobilities presented by such a vessel and its crew.

Conclusion

In this paper, I have introduced three significantly mobile subjects (super-rich, yacht and crew) that warrant more attention from mobilities scholars. I suggest that future studies should account for the further externalities that impact upon and shape super-rich mobilities on board yachts, such as maritime laws and regulations and onshore politics that influence the mobility of yachts. Incorporating super-rich aerial mobilities will also serve to enhance our understanding of the politics of mobility on board as we take in to account the exclusive transportation required in order for the super-rich guests to even arrive at the yacht. In addition the idea of circumscribed mobility can inform future investigations that explore the anchors or the social-spatial politics that restrict or enable the movement of people or things and the effect of doing so. Following Kaufmann (2002) and the notion of motility (potential for mobility), the idea of circumscribed mobility explores the social-spatial politics that although allowing mobility attracts and thus somewhat restricts the movement of others (see also Kesselring 2006). In the case of the super-rich and their use of yachts, their mobility is tied to the desire to perform within a specific geographical area, concentrated around specific port towns in the Cote d'Azur. The movements, routes and journey of the fishing vessel for example are determined by the location and the mobility of shoals, the fisherman returning to his home port after a journey at sea dictated by the direction and collection of fish. Although the fisherman has the tools, skills and experience needed to catch the fish and thus exerts more power, the fish until the moment that they are caught swim free and at will. Cargo ships, with the vessel, its crew and cargo continuously mobilised from port to port yet never free of to roam the high seas arbitrarily despite the vessel and the crew's potential to do so because the crew are incentivised and (legally obliged) to ensure the safe and efficient passage of cargo from shore to shore.

Motive, rhythm and friction as three fundamental aspects of mobility have helped to unravel the complexities of super-rich mobility within the confines of the luxury yacht. Despite the differences in motives, rhythms and experiences of friction, the politics of mobility on board between crew and guests is complimentary rather than conflicting. The super-rich and the crew work together to create and maintain the performance of super-rich mobility – particularly in prominent ports and anchorages where the guests are more likely to be seen. Exploring the politics of super-rich mobility has emphasised the role of performance and showcasing both the vessel and the elite guests. The interactions and performances discussed in this paper are unique therefore to the yacht, yet each vessel with its various cargo whether oil, containers, fish or a member of the Forbes top ten rich list will no doubt have a whole host of social-spatial performances unique to that vessel. Extending these notions of performance and circumscribed mobility and with further utilisation of workplace geographies and related literatures could enable mobilities scholars a platform onto which to explore the social-spatial politics of unusual, or at least non-terrestrial mobile workspaces such as oil rigs, wind farms, lighthouses, submarines, cranes or subterranean spaces.

Acknowledgements

Many thanks go to the two anonymous referees for their insightful and encouraging feedback on a previous draft. Thanks also go to the editors for their invitation to contribute to this special issue. This paper is supported by an ESRC PhD studentship.

References

Adey, P. 2006. "If Mobility is Everything Then It is Nothing: Towards a Relational Politics of (Im) mobilities." *Mobilities* 1 (1): 75–94.

Baerenholdt, J. O. 2013. "Governmobility: The Powers of Mobility." *Mobilities* 8 (1): 20–34.

Beaverstock, J., and J. Faulconbridge. 2013. "Wealth Segmentation and the Mobilities of the Super-rich: A Conceptual Framework." In *Elite Mobilities*, edited by T. Birtchnell and J. Caletrio, 40–61. Abingdon: Routledge.

Birtchnell, T., and J. Caletrio, eds. 2013. *Elite Mobilities*. Abingdon: Routledge.

Blum, H. 2013. "Introduction: Oceanic Studies." *Atlantic Studies* 10 (2): 151–155.

Cidell, J. 2012. "Flows and Pauses in the Urban Logistics Landscape: The Municipal Regulation of Shipping Container Mobilities." *Mobilities* 7 (2): 233–245.

Crang, P. 1994. "It's Showtime: On the Workplace Geographies of Display in a Restaurant in Southeast England." *Environment and Planning D: Society and Space* 12: 675–704.

Cresswell, T. 2010. "Towards a Politics of Mobility." *Environment and Planning D: Society and Space* 28 (1): 17–31.

Edensor, T., and J. Holloway. 2008. "Rhythmanalysing the Coach Tour: The Ring of Kerry, Ireland." *Transactions* 33: 483–501.

Fineman, S. 2003. *Understanding Emotion at Work*. London: Sage.

Hannam, K., M. Sheller, and J. Urry. 2006. "Editorial: Mobilities, Immobilities and Moorings." *Mobilities* 1 (1): 1–22.

Hasty, W., and K. Peters. 2012. "The Ship in Geography and the Geographies of Ships." *Geography Compass* 6 (11): 660–676.

Hay, I. 2013. "Establishing Geographies of the Super-rich: Axes for Analysis of Abundance." In *Geographies of the Super-rich*, edited by I. Hay, 1–25. Cheltenham: Edward Elgar.

Jones, O. 2010. "The Breath of the Moon: The Rhythmic and Affective Time-spaces of UK Tides." In *Geographies of Rhythm*, edited by T. Edensor, 189–203. Oxford: Ashgate.

Kaufmann, V. 2002. *Rethinking Mobility: Contemporary Sociology*. Aldershot: Ashgate.

Kesselring, S. 2006. "Pioneering Mobilities: New Patterns of Movement and Motility in a Mobile World." *Environment and Planning A* 38: 269–279.

Lambert, D., L. Martins, and M. Ogborn. 2006. "Currents, Visions and Voyages: Historical Geographies of the Sea." *Journal of Historical Geography* 32: 479–493.

Lefebvre, H. 2004. *Rhythmanalysis: Space, Time and Everyday Life*. London: Continuum.

Light, L. 1999. "The Floating World of the Rich and Famous." In *Salt beneath the Skin*, edited by T. Duder, 124–132. Auckland: HarperCollins.

Martin, C. 2013. "Shipping Container Mobilities, Seamless Compatibility, and the Global Surface of Logistical Integration." *Environment and Planning A* 45: 1021–1036.

May, J., and N. Thrift, eds. 2001. *Timespace: Geographies of Temporality*, 187–208. London: Routledge.

Peters, K. 2010. "Future Promises for Contemporary Social and Cultural Geographies of the Sea." *Geography Compass* 4 (9): 1260–1272.

Peters, K. 2012. "Manipulating Material Hydro-worlds: Rethinking Human and More-than-human Relationality through Offshore Radio Piracy." *Environment and Planning A* 44 (5): 1241–1254.

Sayer, A., and R. Walker. 1992. *The New Social Economy: Reworking the Division of Labour*. London: Blackwell.

Spence, E. 2014. "Towards a More-than-sea Geography: Exploring the Relational Geographies of Super-rich Mobility between Sea, Superyacht and Shore in the Cote d'Azur." *Area* 46 (2): 203–209.

Steinberg, P. 2013. "Of Other Seas: Metaphors and Materialities in Maritime Regions." *Atlantic Studies* 10 (2): 156–169.

Vannini, P. 2011. "Constellations of (In-)convenience: Disentangling the Assemblages of Canada's West Coast Island Mobilities." *Social and Cultural Geography* 12 (5): 471–492.

Tracking (Im)mobilities at Sea: Ships, Boats and Surveillance Strategies

KIMBERLEY PETERS

Department of Geography and Earth Sciences, Aberystwyth University, Ceredigion, UK

ABSTRACT *This paper explores how national governments exercise regulatory power over spaces beyond their jurisdiction, when activities in those extra-territorial spaces have direct impacts within the boundaries of state concerned. Focusing explicitly on the control of shipping mobilities in the high seas and territorial sea zones, it is contended that apparatus of control, in particular, surveillance, are not only complex across spaces of alternate legal composition and between spaces of national and international law, but also across of the differing conditions and materialities of land, air and sea. Indeed, this paper argues that the immobilisation of the undesirable mobilities of ships and boats is inherently difficult at sea because of its very nature – its mobile legal boundaries, its liquidity compared to 'landed' fixity, and its scale and depth. Drawing on the case study of offshore radio pirates and the tender vessels which travelled ship to shore to supply them with necessary goods, it is reasoned that greater attention must be paid to mobilities at sea in view of forms of governance in this space. The sea is not like the land, or air, legally or materially, and mobilities cannot be governed, controlled and contained in the same ways therefore, as these connected spaces. Thinking seriously about the issues that arise when surveillance of mobilities is taken to sea, can help work towards better understandings for why security at sea proves so problematic and how those issues can be resolved, when the sea is the stage for contemporary geopolitical concerns in the twenty-first century.*

Introduction

> On 28 February 1977 we were informed by the Frontier Police at Calais that the trawler *St Andre des Flandres* registered in the area of Boulogne sur Mer was suspected of delivering provisions to the clandestine broadcasting station Radio Caroline on the ship 'Mi Amigo' anchored in international waters off the English coast. The belief was based on the fact that the boat had loaded up with food...which bore no relation to the size and needs of the crew, which comprised of only three men. On 1 March 1977 the boat left Calais at 9 30

> heading for the English Coast after having filled up with gas oil ... On 2 March 1977 ... A search was carried out ... which found that there was hardly any trace of the provisions and gas-oil fuel taken out the night before. (Home Office HO255/1220)

During the late 1970s, a lone pirate broadcasting vessel, Radio Caroline's *MV Mi Amigo*, was anchored in an area of the English Channel known as Knock Deep, transmitting radio programmes from international waters into the territories of the UK, France and Holland. Anchored fast in position to ensure stability of the enterprise, radio ships were reliant on supplies from the shore to sustain their activities. The above passage recalls one such supply trip and its surveillance by the Radio Regulatory Department of the Home Office (RRD hereafter). Such supplies were chartered from the adjacent shores of England, Holland and France, yet these 'drop offs' of goods necessary to fuel the offshore broadcasting enterprise were illegal following both a pan-European agreement on the eradication of radio piracy (1965) and the UK's Marine &c. Broadcasting Offences Act of 1967 (provision 41.3) which resulted from the aforementioned treaty. The desire to eliminate radio piracy was predicated on the fact that stations lay beyond state control physically, due to their position at sea, but through the mobility of sound waves, were able to intangibly infiltrate the porous state boundary with broadcasts which were tuned into by millions of listeners on shore (see Humphries 2003, 27). That the stations were outside of national control, meant that they could, in effect, broadcast what they wished, in contravention to moral or legal norms within a particular territory. The influence of such stations was a potential threat to state security of the airwaves as 'radio signals ... crossed international borders indiscriminately' (Robertson 1982, 73). In order to ensure security against such a potential breach of order, nation states (unable to exercise power in international waters against such an activity) attempted to control the problem via regulating the elements of the enterprise which fell *within* their legal territorial jurisdiction. This could be achieved by cutting off a vital ship to shore link – tracking the supply boats which provided goods to the radio pirates and intercepting them when they re-entered territorial waters, where, under the jurisdiction of the state they *could* be seized (see Peters 2011).

The above passage illustrates one of the many recordings made by Home Office officials who surveilled the mobilities of supply boats (otherwise known as 'tenders') in an effort to immobilise the broadcasts emanating from international waters. Following this particular interception and 'search', evidence was collated to charge those associated with illegal tendering on the *St Andre des Flandres*. The ring leader, Oonagh Karanjia was fined £500 for orchestrating this particular 'drop' (HO 255/1220). In this paper, drawing on the case study of broadcasting pirate Radio Caroline and its tender operations in the 1970s, I consider how the Labour government secured British territory in view of territorial and extraterritorial mobilities, which had a bearing on life within the boundaries of the nation state. I explore the ways in which a particular method of security – surveillance – was complex when considering its operation across not only differing legal spaces (national and international; the space of the shore, sea and ship), but differing physical or material spaces (the physical composition of the sea compared to and in relation with the land and air). Indeed, examining the surveillance operations led by the RRD and the Essex Constabulary, I demonstrate how the nature of the sea presented particular problems relating to the prohibition of tender boat mobilities, travelling from the pirate radio vessel *Mi Amigo*, to the shore.

Accordingly, this paper unpacks the unique challenges relating to state security when the maritime realm is brought into focus. To date, the study of mobilities and of surveillance has marginalised the sea and ships in investigations. It will be argued, however, that the challenges of surveilling mobilities at sea are different from those on land or air (although these spaces are often connected to the sea through processes of mobility and surveillance) because the sea has a particular legal, fluid and material composition. Whilst this paper focuses on a historical case study of regulation of mobilities at sea to immobilise the aerial vibrations of sound, it is not only past security which may be understood differently through the lens of seas and shipping. In the twenty-first century, territorial security is increasingly played out in the spaces between and beyond national boundaries where there is opportunity to exploit the potential of such zones (see Langewiesche 2004). It is pertinent therefore, to think through the surveillance of shipped mobilities in order to better understand how to govern activities beyond territorial boundaries.

In order to work through these arguments, I split the paper into four parts. I begin by tracing the study of the sea and ships in the social sciences over the past 20 years, accounting for the lack of recognition of this sphere in academic debate. Connected to this, I next consider how scholarship in mobilities studies and surveillance research might be informed through attention to the maritime realm, highlighting recent examples of work that is beginning to fill the watery void in research within these arenas of study. I next introduce radio piracy, the example around which I will explore mobilities and governance of the sea through surveillance practices. Here I outline the numerous (im)mobilities tied up with the enterprise and government strategies of mobilising action and immobilising piracy on the airwaves. I then explore the ways in which the immobilisation of watery pirate radio-related mobilities was a challenge for the British government through three empirically informed sections which draw out the distinctiveness of the sea and ships to the regulatory practices which ensued. Using a variety of data from Home Office records, memos, parliamentary debates and legal documents, I firstly consider mobile legal boundaries at sea, secondly the liquidity of watery spaces, and thirdly the scale and depth of oceans. To finish, I draw conclusions which point towards the broader, contemporary parallels of this discussion that may be reached in order to think seriously about mobility, surveillance and the regulation of shipping in order to secure the sea, and also the land and air.

Seeing the Seas: Mobilising Shipping Research

The sea and shipping occupies something of a paradoxical space in that it is, and has been, simultaneously absent and present in the imagination. As Lavery tells us, the sea is evident all around us, in the tangible goods we have in our homes, which have largely travelled via import channels across the oceans on ships (2005, 359). Ninety-five percent of global trade, on average, is carried by cargo ship; not by air, or across the land (Ibid. 2005, 359). Accordingly, the sea permeates our everyday lives. But this leakage of sea, on to land, via commodity chains is often obscured: the seas and shipping are not associated with the things around us, from table, chairs, electric items or food stuffs. Likewise, the seas, ships and boats have featured in many literary and artistic accounts (see Mack 2011). Here the sea fills the imagination with tales of sea creatures, maritime swashbuckling and sublime imagery, whilst concurrently, as Steinberg notes, being abstracted in Western understanding as an empty void, a mere barrier to cross, for modern and postmodern capitalist flows (2001). Indeed, although in non-Western cultures, the sea has played a much more central social role it has otherwise been predominantly constructed as a space beyond rather than *of* society (Steinberg 2001, 6). Such an abstraction has worked to marginalise the sea in our consciousness, in terms of how it is intimately enfolded with everyday life (Peters 2010). However, since Steinberg's seminal text *The Social Construction of the Ocean* (2001), there has been a steady stream of research that has attended to the widely acknowledged watery void that is evident across social science disciplines, from human geography to sociology (see e.g. Anderson and Peters 2014; Lambert, Martins, and Ogborn 2006; Peters 2010; Steinberg 2010, 2013).

Within this emerging work, the sea has most often been utilised to open up fluid understandings of sociocultural and political phenomena which move beyond constraining 'nation-state centred historical master narratives' (Lambert, Martins, and Ogborn 2006, 480). Other scholars have employed the sea as a metaphoric device for rethinking histories of imperialism and colonialism (see Lambert 2005; Ogborn 2002, 2008) and broader global relations (Linebaugh and Rediker 2000) as the sea represents a space of circulation and exchange. However, critics have argued that such approaches abstract the sea from its 'real' material form. The sea as symbolic, pays attention to what the sea *comes* to symbolise rather than the sea itself – and hence the oceans remain somewhat marginalised in our understandings of them as physical spaces which underpin actual lived realities (Anderson and Peters 2014, 20). Yet with a post-human turn in the discipline, alert to the more-than-human elements that are entwined with our daily lives (see Whatmore 2006), the materiality of the sea, its aesthetic qualities, motion, texture and physical composition are currently driving research agendas (see Jones 2011; Lehman 2013; Peters 2012; Steinberg 2013; Vannini and Taggart 2014). This follows a material turn across the social science disciplines whereby 'matter' is understood as elemental; the building blocks for life; air, fire, water, geology and so on (see Anderson and Wylie 2009). Here the very *nature* of the sea comes into focus as co-composed with human existence (see Lehman 2013; Peters 2012; Vannini and Taggart 2014).

Such work which attends to the connections between the sea as a material reality and the sociocultural life which unfolds in view of these particular watery physical conditions has also brought renewed attention to ships (Peters 2012). The ship has, surprisingly, gained only occasional consideration until recently, in social scientific study (Hasty and Peters 2012, 661). Research concerning ships has focused on vessels as tools in the creation of knowledge about the world (see Hasty 2011; Laloë 2014; Sorrenson 1996), the ship as a material space (assembled and disassembled, see Ryan 2006 and Crang 2010, respectively), and the ship as a site with particular social realities which emerge specifically in view of on board conditions, framed by wider geopolitical contexts (see e.g. Marcus Rediker's *The Slave Ship* where the social canvas of the ship is illustrated in view of broader racial and political constructions of colonialism which foreground the behaviours on board, 2007). However, to date, there has been surprisingly little work on mobilities of ships and shipping and also on the geopolitics of shipping and governance at sea (for exceptions see Cowen 2007, 2012 and Martin 2012).

The lack of attention to the mobilities of ships can be attributed to technological and conceptual shifts over the past century. As Peters explains,

> Today the car, train, plane and virtual networks (the internet for example) are seen as the most important technologies that govern how people move and this view is reflected in the wealth of 'mobilities' research ... The ship, deemed as slow, old-fashioned and out-dated has eluded study. (2010, 1263)

With the recognition that 'our world is a water world' (Anderson and Peters 2014, 1) consideration is now being paid to mobilities facilitated *by* ships (Anim-Addo 2014; Cook and Tolia-Kelly 2010; Stanley 2008) and the mobilities *of* ships (Peters 2012; Hodson and Vannini 2007, see also, the introduction to this issue). Yet there is still, arguably, much work to do in order to take seriously how ships make a world of people, raw materials, services, capital, mobile through global transportation; and how ships themselves are mobile in view of technology, legal barriers and boundaries which separate national space from international space, and due to the very nature of the sea.

Indeed, it is these latter concerns that this paper attends, considering how ships are mobile or immobilised through apparatus of national and international control (focusing specifically on surveillance as an act of regulating space) and in view of the changing physical conditions of the sea at any given time, due to its material composition as liquid, vast and deep. Such a focus also takes the study of surveillance to sea. To date, surveillance studies have both land centric, and also urban centric (see Lyon 2007). In recent years, surveillance studies have expanded to take seriously the surveillance of the air (see Adey 2004a, 2004b) and virtual surveillance of telecommunications and cyberspaces (Lyon 2007). Yet, as Adey notes, the surveillance of port areas, and indeed shipping beyond the shore, in the high seas, has eluded critical examination (2004a, 1367). This is in some respects unsurprising. As Steinberg notes, the sea is, by its nature inaccessible, distant, deep – we cannot study its easily as we might the land; its very physicality providing a barrier to research (Steinberg 1999, 372). It also follows the traditional view of the sea as beyond our daily lives and existence; a space 'out of sight and mind' (Ibid. 1999, 367). Yet, as the sea and ships are now increasingly acknowledged as connected to and integrated with our lived realities, surveillance of these spaces in view of security of both the sea, and the land and air, is necessary in view of twenty-first century concerns: terrorism, environmental degradation and resource exploitation (Langewiesche 2004).

Whilst surveillance studies have paid attention to the ways in which mobilities are tracked, recorded and watched (Bennett and Regan 2004) in light of the multiple, complex ways in which people, goods and ideas are mobile (Sheller and Urry 2006), surveillance of sea-based mobilities have been largely absent (although see Cowen who has explored security of ports 2007, 2010). Yet, what might be learned by considering how surveillance works in watery realms? As Lambert et al. note, the sea has a 'potential' to unlock alternative understandings because, fundamentally, it is a space unlike the land, the air or the virtual sphere (2006, 480). For example, Adey, Whitehead and Williams have theorised the specificity of aerial surveillance in view of the unique perspectives possible from above (2011, 2013). The authors explore the legal shape of air space as jurisdiction alters vertically (Adey, Whitehead and Williams 2011, 177) and further work by Adey, Anderson, and Lobo-Guerrero (2011) explores how the material and elemental quality of air impacts security. Drawing on the volcanic ash cloud of 2010 they demonstrate how the air fails to respect boundaries and movements of ash threatened the security of global transportation flows (2011). In other words, such work has pinpointed the specificity of the legal and material character of air to open up fresh debates regarding mobility and security.

Whilst this paper attends in part to the air (and the movement of sound, ship to shore), here I unpack the specificity of the sea to understanding mobility and surveillance and the surveillance of mobilities. To do so, I take both a legal and a post-human approach, drawing on contemporary theories of the more-than-human to consider the very nature of the sea in the regulative strategies of surveillance that work to immobilise the movements of ships and boats (and on the other hand, consider how those surveilled also harness legal and material conditions to avoid surveillance). To do this, I draw on the case study of offshore radio piracy, which I now introduce in greater detail.

Immobilising Aerial and Watery Mobilities: Introducing Radio Piracy

In Britain, prior to 1964 (when the first radio pirate aiming broadcasts specifically to a British audience began transmissions), there was a monopoly that permitted just three radio stations; the Light, the Third and the Home, all operated by the BBC, to broadcast (Chapman 1992, 31). Due to a government charter (1927) no other broadcasters were allowed to transmit programmes within British territory. Broadcasting was thought to be a powerful medium and consequently the government wanted to ensure it was a resource that was stringently controlled (Lewis and Booth 1989, 52). Having just one agency, the BBC, in charge of broadcasting, guaranteed that what the public consumed through the radio was informing, educative and in good taste (Cain 1992, 12). However, by the 1960s, with a new youth culture emerging, a post-war economic boom, the advent of the portable transistor radio to replace the weighty wireless, and the beginnings of a wave of rock n' roll (see Marwick 1998), there was a demand for an alternative to the BBC's programming, which was often accused of being snobbish and elitist, and failed to air the plethora of new music emerging at the time (Crisell 1997, 27).

Radio pirates realised that they could escape the legislative stranglehold of the British broadcasting monopoly if they were to locate themselves outside of British territory in the high seas zone. From this extraterritorial location, they were physically beyond the reach of British legislation, but the broadcasts they aired could legitimately permeate the boundary back into the state, with no legal consequence. This practice was reliant on the mobility of sound waves moving through the air (Connell and Gibson 2003) – connecting two disparate spaces. Accordingly radio pirate bosses converted ships, destined for international waters, whereby it would be possible to utilise the mobility of signals through the air to transmit popular music to the masses on land. The ships involved in pirate radio endeavours then, were largely immobile (except for the undulation of the vessels, driven by motion of the sea). They would be anchored, often for years at a time, in the same place. Yet from this place they became mobile in a completely different sense. Through the power of broadcasting, the ship moved into every garage, workplace, home, car where the radio broadcasts were picked up by listeners on land.

Following the end of World War II and with the onset of the Cold War, this was perceived as too much of a threat to state security. However, when seeking to regulate Radio Caroline, the government did not aim to control broadcasting frequencies or transmitting output – the factors defined as the problem that required government regulation. Rather, they sought to control the places from which the transmissions originated and which enabled such transmissions – the ships and tender boats at sea (Peters 2013). Actions directed against the air, by way of signal jamming the use of unauthorised or already allocated frequencies would prove fruitless, as Lord Newton described during a Lords' debate in 1964,

> ... it would be costly and it would take some months to arrange; it would also...cause the 'pirates' to keep changing their frequencies in order to overcome the jamming...Jamming stations would be really adding to that situation of uncoordinated and unregulated use of frequencies. (HL Deb 18th June 1964, vol. 258, cc.1380)

The government therefore had to consider how else they might control this aerial problem. In 1966 the government passed the MBO Bill, (which came into effect on August 15th 1967) which would attempt to suppress offshore radio piracy. The Act aimed to achieve the suppression of radio transmissions through making it an offence for a British National to work on a pirate ship in a broadcasting capacity, an offence for boats leaving British docks to supply the ships, or take people to and from the ships and an offence for British companies to advertise the station anywhere inside British territory. Aside from provision c.41.3 the Act was designed to incapacitate the ship to, in turn, to incapacitate broadcasting. The law was in many respects, a law to control all other mobilities which would by default control the aerial mobilities which were the threat to state security. However, when the legislation was brought into force its provisions were not implemented – its mere existence forced the closure of 13 of the 14 pirate radio stations lining Britain's shores in the 1960s (the exception was Radio Caroline who vowed to continue but which also shut down transmissions just 8 months later). In 1974,[1] however, Radio Caroline returned and set sail for an anchorage in the English Channel and for the first time the MBO Act was to be implemented.

The Radio Regulatory Department (RRD) was responsible for the enforcement of the MBO Act, alongside the police and military who also had enforcement powers under the provision of the Bill. For the RRD, it was paramount to regulate the activities of the resurrected radio pirates as they posed the threat of 'noxious' broadcasting. Indeed, in 1970, Radio Caroline temporarily broadcast for two weeks aboard a Dutch pirate vessel in an attempt to sway the general election and oust the Labour government who had legislated again them in 1967. Such broadcasting was demonstrative of the power of these offshore corsairs to reap influence over state subjects. When Radio Caroline returned in a more permanent capacity in 1974, regulation was therefore paramount against the threat to the moral and political security of the state.

Accordingly, the RRD and police kept their eye on movements between the shore and the radio ships – the mobilities of tender boats – which were run by a listening community mobilised into action. If the authorities could watch and record these movements, they could capture and prosecute offenders of the Act, cut off the networks to the ship and control the airwaves. Thus the practice of surveillance at sea, of shipping, was integral to the enforcement of the MBO Act from 1974 to 1980.[2] Although the law gave the authorities other avenues of potential prosecution, such as enforcing the law against British businesses broadcasting advertisements on pirate radio stations, the authorities kept their focus firmly on the ship and its tenders; not these broadcasts (see Peters 2013). However, the immobilisation of tenders was challenging because whilst the act of surveillance in international waters was within the jurisdiction of the British government, they had no power within international waters to incept those mobile tender boats, registered outside of Britain, which facilitated the running of the radio vessel. The authorities would have to wait until those boats re-entered national waters where their law then took precedence. Yet these operations were fraught with difficulty because of the very nature of the sea in terms of both the plural legal spaces involved (national and international; ship, sea and shore) and because of the very material, physical composition and nature of the sea. I next trace these challenging conditions for the regulation of activities at sea in view of state security, beginning with the mobile legal boundaries at sea.

Mobility and Legality

Offshore radio piracy illustrates a case of what Beckmann, Benda-Beckmann, and Griffiths (2006) call 'legal pluralism' where the enterprise operated 'under plural legal constellations' whereby there is a 'coexistence' or 'overlapping' of legal orders which complicate the use of and control of space (2006, 4). In the 1970s, the Radio Caroline vessel *Mi Amigo* was occupying international waters. In this zone, at this time, there was no stipulation in the Law of the Sea (1958) which prohibited broadcasting from vessels or structures in the high seas area (Robertson 1982, 77). As such, within this legal framework, the station was not in breach of international regulation. Moreover, the ship was flagged (conveniently) to Panama, a nation with no law such as the MBO Act, which would prohibit ships to broadcast, as islands of law subject to the state controls of that nation. Subsequently, the ship as portion of Panamanian territory, occupying the international space of the sea, was not in breach of regulations in view of its activities. The Labour government, however, argued that an agreement on frequency allocation from the International Telecommunications Union (ITU hereafter) was breached (1959) (House of Commons Debate 2 June 1964 vol. 695 cc.933). As Lord Aberdare stated in 1964,

> With no regard for … international agreements, Radio Caroline is broadcasting on 197.5 metres and Radio Atlanta on 200.7 metres, neither of them wavelengths allocated to this country. The frequency used by Radio Caroline is close to frequencies in use in Czechoslovakia and Belgium, and the Belgium authorities have already made their protest about interference…. It is therefore essential, if we are to honour international agreements into which we have entered in good faith that we should take urgent steps to close down these "pirate" radio stations. (House of Lords Debate 18 June 1964 vol. 258 cc.1363)

Yet, in a memo responding to parliamentary resistance, a professional lawyer writing on behalf of Radio Caroline objected to the arguments put forward, on the basis that the international Law of the Sea took precedence over the ITU agreement on frequency allocations. As the memo states,

> it is very much part of this memorandum to suggest that an indirect assertion by the States of a power of control over shipping carrying on innocent

> activities on the seas is a direct and real contravention of the policy underlying the High Seas Convention of 1958. (HO255/1007)

Here the Radio Caroline organisation utilised overlapping or fluid boundaries of law to argue their case, knowing that the Law of the Sea carried greater weight because of the historically engrained freedom of ships at sea to carry out activities without interference (for exceptions see United Nations Convention on the Law of the Sea, Article 22, 1958).

Subsequently, in the first instance, Radio Caroline (and other offshore pirates) used the legal nature of the sea specifically to protect their activities, arguing that ships broadcasting 'are doing so on the high seas and in exercise of the undoubted right of freedom of the seas' (Memo, HO255/1007). Resultantly, knowing the long established important of freedom at sea, the British government knew this would have to be respected in any regulation of the activities of the pirates. They could not board and shut down transmissions off the back of the ITU provision – they would have to find indirect methods to contain this extraterritorial problem.

Accordingly, in 1974 when Radio Caroline returned to Knock Deep, the government were forced to enact provisions of the MBO Act. However, the implementation of this law was complex because of the mobility of tender boats and further overlapping legal domains. Indeed, supplying the ship was orchestrated tactically by the Radio Caroline organisation to take advantage of legal plurality (Beckmann, Benda-Beckmann, and Griffiths (2006). In order to evade the provisions of the MBO Act, supplies would be arranged from the shores of France, which was almost equidistant to the anchorage of the *Mi Amigo*, as the British shoreline (HO 255/1219). The mobility of the tendering exercise meant that the Radio Caroline organisation could harness this potential to evade British legal provisions and take advantage of the protection which they were afforded by travelling from French waters to international waters.

However, whilst supplying from France was not illegal (no equivalent of the MBO was in place and broadcasts were not infiltrating French air space and as such were not a French concern) – British officials could ask for assistance from the continent, as evidence collected could be used to prosecute in view of the wider pan-European agreement (1965) on which the MBO Act was based. British ships could not survey in French territorial waters as this was beyond their legal control. They were subsequently reliant on French surveillance. Yet this relied on the French cooperating and surveilling shipping from their shores, of suspected tender boats, flagged conveniently, sailing into international space, on behalf of the British authorities. Such circumstances made surveillance at sea fraught with difficulty because of the number of agencies and legal provisions that came into play. Mobility across legal boundaries by tenders therefore complicated the powers of and possibilities of surveillance at sea by the British. Whilst the British authorities could often conduct their own surveillance of boats moving ship to shore, they were also reliant on the French government for the accumulation of evidence. They could also only intercept tenders within their own national waters, not extraterritorial zones (where it is illegal to board a ship flagged to another nation, Law of the Sea 1958, Article 22) or French waters. This legal plurality, requiring the need for cross-nation coordination, meant that there was only one prosecution against a French national for supplies to the *Mi Amigo* between 1974 and 1980, in spite of numerous launches from Boulogne (HO 255/1219).

Liquidity and Fixity

However, it was not simply the legal canvas of the sea or the fluid mobilities of tenders that complicated regulative strategies of surveillance. The material or physical composition of the sea also played a role. First, the aforementioned borders between territorial and extraterritorial space were rarely clear cut. The sea (in this case) as a liquid element (as opposed to its other states, as a solid in the form of ice), is composed of loose particles. It is the molecular composition of water in liquid form which facilitates its movement, as these looser particulars are then subject to wider elementary forces of wind, gravitational pull and so on (see Jones 2011; Peters 2012). As Chris Bear and Sally Eden ask, in their research of fishery certifications, 'how far can ... strict cartographic boundaries deal with the essential fluidity of seas and oceans?' (2008, 488). In other words, whilst a border or boundary on land, as a solid material, may become in some sense solidified and thus visible and obvious, borders on sea are less 'set' or evident. They are more akin to aerial borders, yet have vertical depth rather than height. Whilst boundary practices on land involve markers of territory (walls, check points and fences); at sea, the materiality of the space dictates somewhat different bordering practices. Whilst the 12 nautical mile territorial sea zone may be clearly marked on a map; a solid line representing a definite distance from the shore separating two legal zones; in reality that line is not solid. No line can be marked across the ocean due to its materiality, no wall can be built, no fence erected, no check point installed. Moreover, with shifting tides back and forth the point of measurement of the territorial sea is constantly changing with the motion of the sea as it ebbs and flows. This materiality and mobility of the sea caused multiple instances of confusion for the RRD because exact legal boundaries could not be clearly identified. This was compounded by the lack of tracking technologies available at the time (the mid to late 1970s).[3]

In 1975, the RRD and Essex Police Constabulary surveilled both the mobility of tenders and of the *Mi Amigo* itself, as it drifted from its permanent anchorage in the Knock Deep channel. In November of that year, a strong Force 8 to 9 wind had whipped up a storm of bad weather and the radio ship had broken its anchor chains and begun to move *in situ* with the mobile sea, its engines having failed (RRD Report, HO255/1219). It was unclear, due to the undistinguishable and imprecise location of the territorial boundary, whether the *Mi Amigo,* and the tenders which came to assist the ailing ship, were in international waters, or the British sea zone. In other words, it was unclear if the ship could be boarded and tenders intercepted (in line with British Law) or whether they were protected from such action by international law (RRD Report, HO255/1219).

On the 13 November, however, broadcasting commenced 'at about 09 30 GMT' from the *Mi Amigo.* 'Navigation equipment ... showed that the broadcast on 1187 kHz was coming from the ship ... its position was considered to be within UK territorial limits' (RRD Report HO255/1219). The word 'considered' used in the report is indicative of the uncertainty over the positioning of the ship and the 'fast twin prop tender' which was supporting the activities as the vessel drifted. Accordingly, believing the ship and tender to be occupying British territory, the RRD and police, who had been surveilling activities from safe distance, boarded the vessels and made several arrests, as well as removing transmitting equipment. When the case came to Southend Magistrates Court in December, however, the surveillance of the RRD and Police was deemed inaccurate. Whilst they believed they were watching a ship in distress and an assisting tender, *within* territorial limits, the moving and ambiguous boundary of the sea could not be determined and as such the 'experts on maritime law advised no further physical or legal measures could be taken against the *Mi Amigo'* (HO255/1219). As such, the surveillance of borders and boundaries at sea is fraught with difficulty because such borders and boundaries are not static, fixed, solid and easily identified – they shift and move as the sea moves, and with only a slight change of position, a vessel may be within or outside of any particular zone of legal jurisdiction, as this instance identifies.

Moreover, such surveillance was made more challenging because of the material liquidity of the sea and how this shaped the technology and means of surveillance possible. Unlike modes of surveillance on land, such as wall-mounted CCTV cameras that work to provide an uninterrupted survey of a given area (Bennett and Regan 2004, 452), such surveillance is not possible at sea because of its materiality. On the one hand, there are few permanent, solid structures at sea from which permanent surveillance could be facilitated. On the other hand, even if there were, such is the vastness of sea spaces; fixity would not necessarily ensure the best vantage point for surveillance. Furthermore, whilst surveillance strategies are varied and not always reliant on fixity (i.e. the CCTV camera may be fixed to the solidity of a wall, but it can pan left to right, up and down), recordings are, by and large, continuous in nature (Lyon 2007, 29). The CCTV camera (unless faulty or vandalised) will pan constantly, providing a continual record of an area. Similarly data surveillance of internet use or telecommunications (via virtual or audio surveillance) may also consist of long term, ongoing records. At sea, such recordings, over such a vast space, are not possible in the same ways. Rather, surveillance is reliant on the mobility of surveillers to travel across the oceans in efforts to capture and record motion, and surveillance is not constant, but occurs as and when necessary in view of a potential threat.

Indeed, in the case of Radio Caroline, the surveillance of tenders was not ongoing or fixed. The RRD and Police used small boats, and in later years helicopters, to survey areas of sea when they believed a tender would be supplying the radio ship. Their presence at sea was not permanent or ongoing, because this was simply not possible on the ocean. Instead, based on intelligence, the RRD and Police had to be ready to sail to sea and watch activities whenever they believed a pick-up/drop-off was occurring. As such, the authorities had to be ready to set sail at short notice, as soon as a tender was detected through forms of audio surveillance (listening to Radio Caroline's broadcasts to detect if tenders were mentioned) and/or visual surveillance (of shorelines). As the Home Office records reveal, they were 'unlikely ... to have more than a bare minimum of warning' about potential tendering and as such, were 'virtually powerless unless (there is) speedy transport available at comparatively short notice' (HO255/1219). As such, the distinct liquid materiality of the sea informed the surveillance measures and practices possible, where the physical composition of the sea eludes fixed and ongoing forms of observation.

Scale and Depth

Furthermore, the wide geographical expanse of open ocean space made it difficult to keep track of all movements, all of the time, especially in an era without satellite technology and with only long-distance photography and notepads as methods of recording (see HO 255 files). Even in contemporary society, with modern Geographical

Positioning Systems (GPS) and 'black boxes', observation at sea remains difficult (the recent example of tracking the missing Air Malaysia flight MH370 in the South China Sea and Indian Ocean is one such example). The surveiller needs to know which ship they are looking for in order to find it, and the naked eye is still relied upon to positively identify the vessel (Langewiesche 2004). In the late 1970s the RRD increasingly relied on aerial surveillance to support sea-based surveillance on boats, as the view from above was broader for watching large areas of ocean and movements of tenders across significant distances (HO 255/1227), (Adey, Whitehead, and Williams 2013). However, even with aerial surveillance, such huge expanses of space were difficult to regulate through strategies of observation. As Mr. Lancefield, of the RRD noted, when on a surveillance mission to identify tenders supplying the *Mi Amigo*, such surveillance of the sea was challenging because of the 'ground' which needed to be covered and the problem of correctly identifying specific ships from an aerial perspective. As the Home Office record states, the helicopter

> was airborne at 12 45hrs … we first searched on a course west of Caroline, but although we spotted quite a few bats (non-supply vessels), we could not identify the tender … we then flew down the Essex coat, but again was unsuccessful. (HO255/1227)

Tracking ships over huge distances because of the size and scale (and even texture) of the sea presents particular challenges in the identification of vessels and the capture of evidence. As the surveillance of ships is predicated on the observation of objects, surveillance at sea was reliant on *seeing* the ships, and therefore being able to detect them in vast open spaces. As the evidence from the Zebra 4 surveillance operation of the RRD reveals, it was difficult to search out tenders even from the air where distances could be traversed more quickly and the perspective was enhanced compared to the horizontal 360-degree viewing from the platform of the surveillance boat. Aerial surveillance using Puma helicopters was an action taken in response to the difficulty of sea level, boat-based surveillance, which was slower and offered a flat horizontal perspective limited to looking across rather than down. Yet by the end of the 1970s, even this approach was not wholly successful (HO 255/1227). Indeed, the government found, during a further pirate radio renaissance in the 1980s, that the only reliable method of immobilising the tenders through regulative surveillance strategies was to watch overtly (a tactic not previously used), in the form of a blockade, known as 'Euro-siege'. This meant positioning surveillance boats within view of the radio ship (Skues 2009, 508). Such surveillance worked to intimidate tenders and curtail their mobilities, preventing them from even approaching the radio ship in the first instance. Such an approach was regarded as threatening in view of the freedom of the seas (see House of Lords Debate 25 July 1990, vol. 521 cc.1547), but gave government agencies the benefit of relative fixity to more reliably track and observe the mobilities of tender vessels, impossible from covert positions or mobile approaches where the tenders were followed back to shore.

Indeed, the journeys back to shore illustrate another way in which surveillance of the mobilities of tenders was problematised because of the materiality of the sea and the seabed. The vertical depth of the sea in the English Channel varies considerably with a number of significant sandbanks altering the sea's movement and the mobilities of ships. When tenders travelled ship to shore they would often utilise the character of the material environment to evade the following surveillance vessels.

Occupying smaller, faster vessels than heavier, larger government boats, tenders could track back into territorial waters on shallower routes, which surveillance vessels could not follow for fear of grounding. Subsequently, the tenders were able to use the speed, size and agility of their boats together with the physical environment at sea – its three-dimensionality and depth – to avoid capture in territorial waters where they could be legally intercepted. Such strategies, boat size, route planning and speed in conjunction with the physical geography of the sea, were used to counter government attempts at regulative control. For example during the Zebra 3 operation (October 1974) a transcription of the surveillance exercise on board the RRD vessel 'The Miranda' reveals the evasive action of the tender to government surveillance.

> Miranda to Robby (code name for Mr. Lancefield of the RRD) 'Tom thinks it is quite possible he (the tender) might be trying to lose us over the sands'
>
> Robby 'Have the shoal (tender) rumbled you, have (they) realised what you are doing, have they taken evasive action over the little sink sands, over?' (HO255/1224)

Likewise, during the Zebra 4 operation (November 1974) the RRD again reported that they were unable to track the mobility of the tender and consequently 'it was decided that the tender must be taking evasive action because we had not spotted it on the MOD Radar, operating from Foulness, and that it was ... slinking up the coast to approach from another direction' (HO255/1227). Utilising depth, and also shelter ('slinking' along the coast) were ways in which the government found immobilisation of tenders challenging in view of the material qualities of the sea. To counter Bennett and Regan's claim that there is 'potentially no hiding' in the 'surveillance of mobilities' (2004, 453), conversely, at sea, as space which appears as a monotonous, flat plain, an open space of exposure and visibility (Levi-Strauss 1973, 338–339), there are examples of 'hiding' through the scale and depth of the seas; its materiality colour, texture, three-dimensionality, size and the mobile technologies and methods of surveillance which are only partially productive in tracking such shipped mobilities.

Moving Forward: Conclusions

Tracing the attempts of the RRD to immobilise the movements of radio pirate suppliers, I have demonstrated how overlapping legal spheres between national and international space and between laws of the air and sea, impacted the possibilities of 'where' surveillance could occur and how surveillance could be enacted. I have further demonstrated how the complex material physicality of the sea as a liquid space without the benefits of fixity for continuous surveillance practices, and the sea as a space of vast proportions and depth, impacted the ability for government missions to follow through observational operations successfully. Surveillance was often patchy and tender vessels could evade detection in spite of the open plateau of space occupied. Indeed, the open ocean, far from being a space of easy visuality, had a texture, depth and vastness which meant recognition of tenders was difficult, most notably from the deck of the surveillance boat, but also from the improved vantage point of the air. Subsequently, it is important to think seriously about the material and changing characteristics of the sea and how these shape regulative practices – not because these conditions can be changed, but so as to think about manipulative strategies for working around these conditions in light of state security (see also Peters 2012).

The case study presented here is a historic one and the conclusions drawn from this are, resultantly, shaped by a consideration of the Law of the Sea in place at the time (1958), technological methods of surveillance, and the broader socio-political climate of the time. Indeed, in 1982 the Law of the Sea changed to revoke rights to broadcast at sea (Article 22, 1982) and technological methods of surveillance have since improved to include GPS and infrared monitoring. The case study of Radio Caroline is also one tied up to post war change and the moral liberalisation of society (Marwick 1998) which drove state desires to protect shores from 'noxious' broadcasts that, unlike the BBC's output, might not be in good taste. However, some broader points of consideration can still be drawn for thinking about the sea in contemporary times, as a space connected to land and air spaces, and therefore, a space relevant for unpacking the regulation of mobilities which may directly or indirectly move beyond the sea to these cognate spheres.

In the documentary 'Royal Navy Caribbean Patrol' (2011), television cameras follow the daily lives of sailors whose role it is to secure the seas and shores of maritime threats – threats that are not simply oceanic insecurities, but that have broader influences on landed and aerial life. In Episode One, the scene is set for the series, introducing the vessel HMS Manchester, which will undertake a 7-month deployment to survey international waters adjacent to Montserrat, tracking drug smugglers. Drug smuggling is an example of the ways in which activities at sea (as such pirate radio) cannot be disconnected from connected spaces, when supplies of drugs find their way onto land, and into bars, clubs and the pockets of dealers. Similar to events surrounding offshore broadcasting piracy, engaging with smuggling at sea is a method of securing the shore before the shore is even reached. Smuggling represents a current global concern, a threat to order in numerous states as supplies cross territorial boundaries when the sea is utilised as a crossing zone for trade. However, as the blog postings on the internet web page for 'Caribbean Patrol' reveal, there is contention over the Royal Navy's role in the Caribbean Sea, given most supplies from this area arrive on US rather than UK shores (2011). As one posting remarks, 'America's drug problem is America's problem... I wonder how much this tour cost and would it have been better spent combatting piracy off the horn of Africa?' (Peters 2011, as quoted in Royal Navy Caribbean Patrol). The sea is a place where surveillance as a regulative strategy becomes contentious according to overlapping national concerns. The documentary further shows the care taken to not intercept smugglers until the crew confirm that the activities and people they observe are in breach of the Law of the Sea (1982), so as not to illegally board vessels, in a space of legal plurality. Moreover, the documentary demonstrates the harsh realities of surveillance in the Caribbean during hurricane season when the material qualities of the sea alter from a calm surface ideal for tracking possible smuggling vessels with radar sensors and then binoculars, to the issues of grappling with such surveillance in challenging circumstances where the sea's physical liquidity causes the ship to list and objects of surveillance to be lost from sight (2012). In these instances, the HMS Manchester relies on specific tactics to counter the physical qualities of the sea. Communications with land; the use smaller inflatable craft which can move through the water at speed; and the deployment of aerial surveillance track suspect vessels when the shipboard horizontal view is obscured (2012). Such strategies and approaches to surveillance are markedly different from surveillance on land, where firstly, legal boundaries are (mostly) clear cut in view of national jurisdiction; and secondly, the materiality of the land lends itself to alternative practices of surveillance, specific to the elemental solidity of a given area. Such strategies also differ to those employed in skies, due to the depth, colour and mobility of the sea as a liquid element. In short, the distinct qualities of the sea mean that the Royal Navy prepares for surveillance in view of these specific legal and physical conditions, adapting practice accordingly.

Such adaptation and a greater understanding of the possible legal and material conditions which the sea presents, are fundamental not just in re-visioning how we understand a past phenomenon (in this case the security of land and air spaces from offshore radio piracy), but are also vital for thinking seriously about threats to the security of sea, land and air in future. Attention to the sea is paramount with growing acknowledgement that what happens at sea is enfolded with what happened on land or in the air (Anderson and Peters 2014). As Langewiesche rather pessimistically argues, the ocean is at the heart of twenty-first century concerns over state security because,

> [g]eographically, it is not the exception to our planet, but by far its greatest defining feature. By political and social measures it is important too – not merely as a wilderness that has always existed or as a reminder of the world as it was before, but also quite possible as a harbinger of a larger chaos to come. (2004, 1)

This paper has gone some way in introducing some of the considerations that should be taken into account that are different at sea, to the land, which must be recognised in acts of regulation in order to secure the sea, land and air, whilst also adding to literatures on mobilities and surveillance which have yet to fully 'go to sea'. Yet there is more work to do to think through the range of mobilities at sea which may be subject to surveillance. Here I have focused on ships and boats as a much maligned focus in mobilities and surveillance literature, but this research must also expand to think not only about the means of mobilities at sea, but the mobilities of things carried by ship, and their surveillance (i.e. containers, see Cowen 2010; Martin 2012) and surveillance under the surface of the sea in the form of submarine surveillances. However, in this paper, I have contributed to timely debates which cross-cut and bring together the study of the ocean with mobilities and surveillance studies, arguing that methods of immobilising undesirable mobilities in efforts to secure state concerns is complex when we take the particularities of the sea – legally and materially – into account.

Acknowledgements

I appreciate the comments of the anonymous reviewers, and those of Dr Rhys Dafydd Jones who read an earlier version of this paper. These thoughtful additions have helped greatly in sharpening the argument.

Notes

1. Radio Caroline in fact returned in 1972 alongside Dutch pirate vessels Radio Northsea International (RNI) and Radio Veronica, protected because Holland had yet to enact a version of the MBO Act.

2. In 1980 the *Mi Amigo* sank.
3. Nowadays there are GPS satellite technologies that allow ships and ports to more accurately plot and map locations.

References

Adey, P. 2004a. "Surveillance at the Airport: Surveilling Mobility/Mobilising Surveillance." *Environment and Planning A* 36: 1365–1380.

Adey, P. 2004b. "Secured and Sorted Mobilities: Examples from the Airport." *Surveillance and Society* 1 (4): 500–519.

Adey, P., B. Anderson, and L. L. Guerrero. 2011. "An Ash Cloud, Airspace and Environmental Threat." *Transactions of the Institute of British Geographers* 36 (3): 338–343.

Adey, P., M. Whitehead, and A. Williams. 2011. "Air-Target: Distance, Reach and the Politics of Verticality." *Theory, Culture and Society* 28 (7–8): 173–187.

Adey, P., M. Whitehead, and A. Williams, eds. 2013. *From Above: War, Violence and Verticality*. London: Hurst & Co.

Anderson, J., and K. Peters, eds. 2014. *Waterworlds: Human Geographies of the Ocean*. Farnham: Ashgate.

Anderson, B., and J. Wylie. 2009. "On Geography and Materiality." *Environment and Planning A* 41 (2): 318–335.

Anim-Addo, A. 2014. "With Perfect Regularity Throughout': More-Than-Human Geographies of the Royal Mail Steam Packet Company." In *Waterworlds: Human Geographies of the Ocean*, edited by J. Anderson and K. Peters, 163–176. Farnham: Ashgate.

Bear, C., and S. Eden. 2008. "Making Space for Fish: the Regional, Networked and Fluid Spaces of Fisheries Certification." *Social and Cultural Geography* 9 (5): 487–504.

Benda-Beckmann, F. V., K. V. Benda-Beckmann, and A. Griffiths. 2006. *Spatializing Law*. Farnham: Ashgate [online]. Accessed January 18 2010. www.ashgate.com/pdf/SamplePages/Spatializing_Law_Ch1.pdf

Bennett, C. J., and P. M. Regan. 2004. "Surveillance and Mobilities." *Surveillance and Society* 1 (4): 449–455.

Cain, J. 1992. *The BBC: 70 Years of Broadcasting*. London: British Broadcasting Corporation.

Chapman, R. 1992. *Selling the Sixties: The Pirates and Pop Music Radio*. London: Routledge.

Connell, J., and C. Gibson. 2003. *Sound Tracks: Popular Music, Identity and Place*. London: Routledge.

Cook, I., and D. Tolia-Kelly. 2010. "Material Geographies." In *The Oxford Handbook of Material Culture Studies*, edited by D. Hicks and M. Beaudry, 99–122. Oxford: Oxford University Press.

Council of Europe. 1965. "European Agreement for the Prevention of Broadcasts Transmitted from Stations Outside of National Territories." European Treaty Series – No. 53. Strasbourg [online]. Accessed January 16 2008. http://conventions.coe.int/Treaty/EN/treaties/Html/053.htm

Cowen, D. 2007. "Struggling with Security: National Security and Labour in the Ports." *Just Labour* 10: 30–44.

Cowen, D. 2010. "A Geography of Logistics: market Authority and the Security of Supply Chains." *Annals of the Association of American Geographers* 100 (3): 600–620.

Crang, M. 2010. "The Death of Great Ships: Photography, Politics, and Waste in the Global Imaginary." *Environment and Planning A* 42 (5): 1084–1102.

Crisell, A. 1997. *An Introductory History of British Broadcasting*. London: Routledge.

Hansard. House of Commons (HC). Debate 2nd June 1964, vol. 695 cc.933 [online]. Accessed September 28–30 2009. http://hansard.millbanksystems.com/commons/1964/jun/02/pirate-radio-ships-and-local-sound-1

Hansard. House of Lords (HL). Debate 18th June 1964, vol. 258 cc.1363-82 [online]. Accessed September 28–30 2009. http://hansard.millbanksystems.com/lords/1964/jun/18/pirate-broadcasting

Hansard. House of Lords (HL). Debate 25th July 1990, vol. 521 cc.1547 [online]. Accessed September 28–30 2009. http://hansard.millbanksystems.com/lords/1990/jul/25/broadcasting-bill-2

Hasty, W. 2011. "Piracy and the Production of Knowledge in the Travels of William Dampier, c.1679–1688." *Journal of Historical Geography* 37 (1): 40–54.

Hasty, W., and K. Peters. 2012. "The Ship in Geography and the Geographies of Ships." *Geography Compass* 6 (11): 660–676.

Hodson, J., and P. Vannini. 2007. "Island Time: The Media Logic and Ritual of Ferry Commuting on Gabriola Island, BC." *Canadian Journal of Communication* 32: 261–275.

Humphries, R. C. 2003. *Radio Caroline: The Pirate Years*. Yately: The Oakwood Press.
Jones, O. 2011. "Lunar-solar Rhythmpatterns: Towards the Material Cultures of Tides." *Environment and Planning A* 43: 2285–2303.
Laloë, A. 2014. "Plenty of Weeds and Penguins': Charting Ocean Knowledge." In *Waterworlds: Human Geographies of the Ocean*, edited by J. Anderson and K. Peters, 39–50. Farnham: Ashgate.
Lambert, D. 2005. "The Counter-revolutionary Atlantic: White West Indian Petitions and Proslavery Networks." *Social and Cultural Geography* 6 (3): 405–420.
Lambert, D., L. Martins, and M. Ogborn. 2006. "Currents, Visions and Voyages: Historical Geographies of the Sea." *Journal of Historical Geography* 32: 479–493.
Langewiesche, W. 2004. *The Outlaw Sea: A World of Freedom, Chaos, and Crime*. London: Granta Books.
Lavery, B. 2005. *Ship: 5000 Years of Maritime Adventure*. London: Dorling Kindersley.
Lehman, J. S. 2013. "Relating to the Sea: Enlivening the Ocean as an Actor in Eastern Sri-Lanka." *Environment and Planning D: Society and Space* 31: 485–501.
Levi-Strauss, C. 1973. *Tristes Tropiques*. London: Jonathan Cape.
Lewis, P. M., and J. Booth. 1989. *The Invisible Medium: Public, Commercial and Community Radio*. Basingstoke: Macmillan.
Linebaugh, P., and M. Rediker. 2000. *The Many Headed Hydra: The Hidden History of the Revolutionary Atlantic*. London: Verso.
Lyon, D. 2007. *Surveillance Studies: An Overview*. Cambridge: Polity Press.
Mack, J. 2011. *The Sea: A Cultural History*. London: Reaktion.
Marine &c., Broadcasting (Offences) Act (c.41). 1967 [online]. Accessed April 15 2008. http://www.statutelaw.gov.uk/content.aspx?LegType=All+Primary&PageNumber=61&NavFrom=2&parentActiveTextDocId=1185362&ActiveTextDocId=1185362&filesize=86262
Martin, C. 2012. "Containing (Dis)order: A Cultural Geography of Distributive Space." Unpublished Thesis. Royal Holloway University of London.
Marwick, A. 1998. *The Sixties: Cultural Revolution in Britain, France, Italy and the United States 1958–1974*. Oxford: Oxford University Press.
The National Archives NA HO255/1007. Date range:1964. Home Office and predecessors: Radio Regulatory Department Registered Files.
The National Archives NA HO255/1219. Date range: 1974–1979. Home Office and predecessors: Radio Regulatory Department Registered Files.
The National Archives NA HO255/1220. Date range: 1976–1977. Home Office and predecessors: Radio Regulatory Department Registered Files.
The National Archives NA HO255/1224. Date range: 1973–1975. Home Office and predecessors: Radio Regulatory Department Registered Files.
The National Archives NA HO255/1227. Date range: 1970–1975. Home Office and predecessors: Radio Regulatory Department Registered Files.
Ogborn, M. 2002. "Writing Travels: Power, Knowledge and Ritual on the English East India Company's Early Voyages." *Transactions of the Institute of British Geographers* 27 (2): 155–171.
Ogborn, M. 2008. *Global Lives: Britain and the World 1550–1800*. Cambridge: Cambridge University Press.
Peters, K. 2010. "Future Promises for Contemporary Social and Cultural Geographies of the Sea." *Geography Compass* 4 (9): 1260–1272.
Peters, K. 2011. "Sinking the Radio Pirates: Exploring British strategies of Governance in the North Sea, 1964–1991." *Area* 43 (3): 281–287.
Peters, K. 2012. "Manipulating Material Hydro-worlds: Rethinking Human and More-than-human Relationality through Offshore Radio Piracy." *Environment and Planning A* 44 (5): 1241–1254.
Peters, K. 2013. "Regulating the radio pirates: Rethinking the Control of Offshore Broadcasting Stations through a Maritime Perspective." *Media History* 19 (3): 337–353.
Rediker, M. 2007. *The Slave Ship: A Human History*. London: John Murray.
Robertson, H. B. 1982. "The Suppression of Pirate Broadcasting: a Test Case of the International System of Control of Activities Outside National Territory." *Law and Contemporary Problems* 45 (1): 71–101.
Royal Navy Caribbean Patrol (TV Documentary). 2011. Accessed January 16 2012. http://www.channel5.com/shows/royal-navy-caribbean-patrol
Ryan, J. R. 2006. "'Our Home on the Ocean': Lady Brassey and the Voyages of the Sunbeam, 1874–1887." *Journal of Historical Geography* 32: 579–604.

Sheller, M., and J. Urry. 2006. "The New Mobilities Paradigm." *Environment and Planning A* 38: 207–226.

Skues, K. 2009. *Pop Went the Pirates II*. Norfolk, VA: Lambs' Meadow.

Sorrenson, R. 1996. "The Ship as a Scientific Instrument in the Eighteenth Century." *Osiris* 11: 221–236.

Stanley, J. 2008. "Co-venturing Consumers 'Travel Back': Ships' Stewardesses and Their Female Passengers, 1919–55." *Mobilities* 3 (3): 437–454.

Steinberg, P. E. 1999. "Navigating to Multiple Horizons: Towards a Geography of Ocean Space." *The Professional Geographer* 51 (3): 366–375.

Steinberg, P. E. 2001. *The Social Construction of the Ocean*. Cambridge: Cambridge University Press.

Steinberg, P. E. 2010. "The Deepwater Horizon, the *Mavi Marmara*, and the Dynamic Zonation of Ocean Space." *The Geographical Journal* 177 (1): 12–16.

Steinberg, P. E. 2013. "Of Other Seas: Metaphors and Materialities in Maritime Regions." *Atlantic Studies* 10 (2): 156–169.

United Nations Convention on the High Seas. 1958 [online]. Accessed January 16 2008. http://untreaty.un.org/ilc/texts/instruments/english/conventions/8_1_1958_high_seas.pdf

United Nations Convention on the Law of the Sea. 1982 [online]. Accessed January 16 2008. http://www.un.org/Depts/los/convention_agreements/texts/unclos/unclos_e.pdf

United Nations International Telecommunication Convention. 1959 [online]. Accessed January 17 2008. http://www.itu.int/dms_pub/itu-s/oth/02/01/S020100001C4002PDFE.pdf

Vannini, P., and J. Taggart. 2014. "The Day We Drove on the Ocean (and Lived to Tell the Tale about it): Of Deltas, Ice Roads, Waterscapes and Other Meshworks." In *Waterworlds: Human Geographies of the Ocean*, edited by J. Anderson and K. Peters, 89–102. Farnham: Ashgate.

Whatmore, S. 2006. "Materialist Returns: Practising Cultural Geography in and for a More-than-human World." *Cultural Geographies* 13: 600–609.

The Packaging of Efficiency in the Development of the Intermodal Shipping Container

CRAIG MARTIN

School of Design, University of Edinburgh, Edinburgh, UK

ABSTRACT *This paper addresses different forms of spatio-temporal ordering in the stowage and handling of cargo on board cargo vessels, as well as docksides. Whilst the introduction of containerisation profoundly altered the urban geographies of many large port cities, as well as devastating the communities built around maritime labour, the core argument developed in this paper concerns the incremental development of spatio-temporal ordering strategies and practices. In particular, it situates the intermodal shipping container within a trajectory reaching back to earlier forms of unitisation such as crates and pallets. In doing so the paper outlines a genealogy of* packaged efficiencies, *arguing that the central thread linking maritime cargo handling practices in the twentieth century is the unitisation of shape. However, it concludes that the intermodal container achieved global hegemony through the packaged* systemic *efficiencies of standardisation.*

Introduction

A recent report in *The Guardian* newspaper (Neate 2013) discussed the increasing size of contemporary container ship designs, with the newest Triple-E Class vessels capable of carrying up to 18,000 TEU's (or twenty-foot equivalent unit containers) at any one time. Although one has to be cognisant of the desire to overtly fetishise such figures (especially the tendency for equivalencies: for example, Triple-E's will be able to carry containers that would fill a train 68 miles long (Neate 2013)), these are important registers both of the continued global strategic power of the container shipping industry, as well as the technological sophistication of such maritime architecture. For example, their scale is likely to have significant effects on the strategic geopower of specific container ports throughout the world: at present no North or South American ports have the capacity to handle vessels of this scale (Neate 2013).

Just as the move to containerisation in the 1970s profoundly altered traditional maritime port communities (Port of London Authority 1979), and the scale of post-Panamax vessels shifted the operation of the Panama canal as a container vessel trade route (McCalla 1999), so the increasing scale of the largest container ships has the potential to reconfigure the significance of ports capable of handling these vessels, with China–Europe becoming the key route for this scale of container ship (see Airriess 2001). This story also highlights an important aspect of the relationship between commercial ships and the geographies of globalisation (Dicken 2011). Whilst this has a long historical trajectory (Gilroy 1993; Steinberg 2001; Law 2003; Parker 2010) it is rather telling that as of early 2011 there was a global fleet of nearly 5000 container ships, carrying an equivalent of 14 million containers (Institute of Shipping Economics and Logistics 2011, 5). Above all these figures provide a stark reminder of just how powerful the container shipping industry is both in terms of the geographies of the sea, but also in so many other areas of contemporary society such as retail (Wrigley 2000), manufacturing technologies (Dicken 2011, 181), and geosecurity (Cowen 2010a, 2010b; Martin 2011), to name just three. Indeed for David Harvey the development of intermodal containerisation was 'one of the great innovations without which we would not have had globalisation, [or] the deindustrialisation of America' (Harvey, cited in Buchloh, Harvey, and Sekula 2011; also see Shaw and Sidaway 2010, 509).

The historical advance of containerisation, whilst not widely disseminated (see Teräs 2007, 138), has recently garnered a growing amount of attention from within the academic community, some more critically attuned than others (Jackson 1983, 154–155; Hunter 1993; Broeze 2002; Cudahy 2006; Levinson 2006; Cidell 2012; Martin 2012). A key objective of this paper is to demonstrate how these taken-for-granted objects and their attendant handling processes and wider mobilities stand as a marker of decidedly complex articulations of the spatial dynamics of contemporary capitalism (Sekula 1996). In particular, the mobility of containers (Cidell 2012) on a global scale is determined by the infrastructural power of intermodal containerisation, defined as an integrated transport system that makes it commercially feasible to deliver goods door-to-door (Talley 2000, 933–934). Containerisation and its foundations in the systems-driven development of nascent consumer capitalism in the 1950s (Forrester 1958; see Gomes and Mentzer 1988; DeLanda 1991; Allen 1997, 110; Easterling 1999a) offers us a useful vantage point from which to consider the origins of such spatio-temporal ideologies, and how they have grown into a wider assemblage of logistics and supply chain power (Toscano 2011; Neilson 2012; Mezzadra and Neilson 2013; Kanngieser, 2013).

In this paper, I aim to locate the development of the shipping container with specific regard to the politics of spatio-temporal organisation in the context of shipping mobilities, and the wider geography of ships (Hasty and Peters 2012). As the intermodal qualities of containerisation demonstrate, the overt focus is not with the ship as such, but rather the role of the intermodal container as a bridging device that crosses a variety of transportation interfaces. A fundamental objective of this paper is to argue that the intermodal container emanates out of a genealogy of unitisation within the wider historical geographies of transportation. This genealogical approach (Foucault 1977; Gutting 1990) is utilised in order to identify the forms of organisational logic present within the packing, stowing and handling of pre-containerised, or break-bulk cargo. In this context I develop my core argument concerning the packaging of efficiency: that is, firstly the move towards the spatial regularity of cubic packing, and secondly the consolidation of these spatial motifs into the total redesign of the cargo handling and transportation system.

The arguments proceed as follows. The paper begins by using two UK Government-commissioned reports as a backdrop to wider debates on the move towards intermodal containerisation in the 1950s and 60s, and particularly the shifts in the organisation of space-time, as well as more subtle material practices and procedures. Whilst the impact on the urban and economic geographies of cities throughout the world cannot be underplayed (Goldblatt and Hagel 1963; Banham 1967; Smith 1989; Sekula 2000; Harvey 2010, 16), the main remit of this section is to focus on the context of traditional pre-intermodal forms of cargo packing and handling. It is also to argue that the supposed paradigm shift that containerisation embodied, or the so-called 'container revolution' (Schmeltzer and Peavy 1970), was actually much more gradual. Prior to the development of full-scale intermodal containerisation in the late 1960s, spatial efficiencies in the form of cargo handling and ordering was present. Where, perhaps, these practices differ from intermodal containerisation was in the *localised* approaches to modes of spatial ordering. The key conceptual foundation of this argument comes from Pye's (1964) notion of shape-determining systems, and I utilise this to differentiate between localised means of spatio-temporal organisation in the stowage of break-bulk cargo and the move towards universalised forms associated with unitisation, mechanisation and automation.

The next section focuses on the development of unitised forms of cargo packaging in the form of pallets, crates and particularly non-standardised containers as embryonic versions of the later fully standardised intermodal containers. Setting these within a broader techno-aesthetic context of regularisation, unitisation and serial logic (Giedion 1948; Sekula 1996; Easterling 1999b) it is argued that *nascent* forms of packaged efficiency were present with these early non-standardised containers, particularly with regard to these containers as enclosed spatial units exhibiting uniformity of shape and thus some of the spatial efficiencies of the intermodal container (Bohlman 2001).

The section following this addresses the extension of packaged efficiencies into what I term totalised or systematised, packaged efficiencies. So whilst the utilisation of cubic space in the design of non-standardised containers provided one form of packaged efficiency, the wider packaging of the entire freight transport system was not evident. The spatial homogenisation of cargo shape represented a developing sense of organisational logic, but up until this point in the late 1960s it also confirmed the inconsistencies in the attendant infrastructure of cargo handling and mobility.

This differs with intermodalism where we see the distribution of the same logic of efficiency and control across the entire transport infrastructure, embracing both the shipping industry, as well as the road and rail freight sectors. Following the wider discussions concerning orderings in the previous sections the conceptual bedrock of this section is the issue of stabilisation through standardisation. I look to the attempts to stabilise the interconnections between the various elements of the cargo handling system, in part to consider the notion of controlling inter-changeability through the delegation of effort from localised to universal procedures. This bears on my central idea of totalised packaging of efficiency, where the improvised tactics of early modes of spatial organisation are consolidated into a tightly coupled apparatus of standardised and universalised approaches.

Localised Ordering in Pre-containerised Cargo Handling

> The traditional modus operandi of the world's trade really is outmoded. (Gunston 1968, 62)

Writing in the journal *Design* in 1968, Bill Gunston made the case for wide-scale implementation of a standardised global freight transportation system: that of containerisation. For him the development of an integrated system would reflect the wider notion of Modern technological progress developing at this time (see Ellul 1964). Gunston described the inappropriateness of 'a sweating army of stevedores' loading loose cargoes of break-bulk items such as bales, sacks or dented boxes onto technologically sophisticated vehicles like the then-new Boeing 747 (Gunston 1968, 62). Instead of such incongruities Gunston's article was inflected with idealised images of a system that operated efficiently: where the inconsistencies of commodity shape and form were smoothed out through the unifying force of homogenised, unitised and standardised containers; where there was a seamless transition across various transport interfaces; where containers moved off ships onto waiting trucks or trains; and ultimately of a system that mirrored the technologically determined consumer landscape of the period (see Banham 1967).

Such sentiments echoed other contemporaneous arguments extolling the benefits of a universally recognised system of freight distribution ("Uniform Containerization of Freight" 1969; "Container Transporter Crane" 1970). In the UK, a report commissioned in 1966 by the British Board of Trade from the management consultants McKinsey & Company outlined the potential benefits of containerisation for British trade, and the global economy (McKinsey & Co. 1966). This initial report was followed by a more extensive one in 1967, *Containerization: The Key to Low-cost Transport* (McKinsey & Co. 1967). Central to both was the relationship between standardisation, infrastructural inter-linkage, spatio-temporal efficiency, and lowered transport costs. Four main conclusions from the move to standardised containerisation were identified:

(1) A reduction in transport costs.
(2) Larger economies of scale which become possible with larger container ships.
(3) Integration and consolidation of the transport industry.
(4) Growing importance of transport for global trade (McKinsey & Co. 1967, iv).

The imperative of containerisation at this time was exemplified by the reduction in overall costs due to lower packaging and transportation costs, coupled with the increased speed of transit: allegedly reduced, for example, from 20 to 10 days on a journey from Birmingham, UK to Chicago (McKinsey & Co. 1966, 2). Further to this, the losses from theft and damage seen with break-bulk cargoes would be reduced, and thus resultant insurance costs. A reduction in cargo handling costs at ports would be a result of the increased speed of handling and a significant reduction in labour costs. Perhaps, most starkly it was argued that reduction in costs could be accomplished by automation and systematisation, leading to a situation where it may have been possible to 'eliminate most of the labour formerly needed' (McKinsey & Co. 1967, 4).

The core arguments in favour of containerisation were that the system of freight distribution up to this point had been typified by excessive labour costs, inefficient means of cargo handling on the dockside and on board ships, as well as a lack of integration across the various freight handling sectors. This resulted in a situation where 'manufacturers, road hauliers, freight forwarders, shippers, shipping companies, railways, stevedoring companies, consolidators' operated in separation to one another (Gunston 1968, 59). This lack of consolidation and fragmentation of the sector was also said to be reflected in the inefficiencies of handling cargo (Broeze 2002, 9). The moments of interchange of cargo between the ships, warehouses, packing crates, cranes, trucks, etc. were incompatible, resulting in the need to load and unload between each. The loading and unloading of ships was a lengthy, time-consuming process: in one example from 1938 individual lamb carcasses are shown being discharged from the *Clan MacDougall* at the port of Tilbury ("Clan MacDougall" 1938). Stevedores on board the vessel, unpacked the carcasses and loaded them individually onto netting, for the crane to then winch them onto the dockside before being packed in the dockside warehouse ready to be collected by a road haulage company at a later time. Such scenes point to the time and effort necessary to handle cargo, with each carcass individually wrapped and treated as a singular entity. Labour-intensive cargo handling procedures were, in part, the result of the sheer diversity of break-bulk or general items which were typified by a lack of uniformity: disparate arrays of packaged items such as crates, pallets, individual cans, planks of timber, barrels, sacks, bales of wool, boxes of fruit, hogsheads of tobacco, as well as animal carcasses, were prevalent (Cargoes 1940). The handling of cargo itself relied upon large numbers of stevedores (or longshoremen in the USA) to both lift these individual items, manoeuvre them onto winches and to arrange them in the holds of the cargo vessels, using the ubiquitous stevedore hooks and other improvised means such as sacking placed around footgear to prevent stevedores from slipping in the hold ("Stevedore Foot Sacking" 1952; also see Goldblatt and Hagel 1963, 13).

The complexities of cargo handling techniques were most evident in the packing of break-bulk goods in ships' holds. Prior to containerisation the job of the stevedore was to counter the spatial limitations of individual cargo shape by packing the hold as efficiently as possible, in order to 'get the most possible goods into the box' (Ford 1950, 60), or what was commonly known as the 'tight stow' (Goldblatt and Hagel 1963, 30). Decisively, the task of the stevedore prefigured the spatio-temporal logic of containerisation as they too were charged with ordering the space of the ships' hold. If we take the case of stowing barrels in the ships' hold their shape necessitated the use of specific methods to secure the barrels, but more importantly to counter the limitations of their shape and pack as many barrels into the hold (Ford 1950, 83). The shape of the barrels necessitated the use of heavy dunnage – pieces of wood (often cordwood) that secured and filled-in the inconsistencies of the barrel shape as they were laid on their sides on top of one another, whilst also alleviating the movement of cargo during voyage. Dunnage also referred to simple wooden boards nailed together to create a framework to protect break-bulk cargoes and regularise the load.[1] Loose items of cargo such as sacks would be stacked in the hold and dunnage boards constructed to form a fence-like structure to hold the sacks in place. In effect the role of dunnage was to create an elementary form of regularity in order to facilitate the packing of the hold. This use of dunnage suggests a level of improvisation, whereby the nailing of simple wooden boards was the most practical method of countering the sheer diversity of cargo items. Overall this demonstrates the level of effort required to literally stabilise the load in the hold of the ship. It also illustrates a further form of stabilisation at the conceptual level. For such modes of ordering, whilst somewhat rudimentary in comparison to the calculative logic of contemporary computer-controlled loading (Bratton 2006; Corry and Kozan 2008), were examples of early forms of spatio-temporal organisation designed to overcome and regularise the diversity of shape. This is perhaps most evident with cargo planning diagrams. Cargo plans and hatch lists enabled ships' holds to be loaded according to the correct weight distribution, a critical factor in enabling vessels to sail safely without listing. Further to this, the correct distribution of cargo in the hold facilitated the most efficient and quick off-loading at a ship's various ports of call (see Branch 2007, 245). Finally, the loading patterns of the hold meant that the correct number of stevedores could be employed to unload the cargo (Ford 1950, 205). Such procedures testify to a significant form of spatial knowledge and awareness, both on a practical level, and tellingly with regard to the economics of loading and unloading. As an unfolding marker of late Modernity the question of speed was central (see Tomlinson 2007). The remit of cargo stowage on board the vessels was to 'to arrange the stowage so that the speed of loading and unloading is at a maximum and the cost to a minimum' (Ford 1950, 60). The length of time taken to achieve this level of coordination in comparison with containerisation (five and a half days for traditional handling, 40 hours for an early container ship (Goldblatt and Hagel 1963, 108)) should not obscure the efficiency and skill of loading and discharging ships.

However, for Gunston (1968), and others advocating the move toward containerisation at this time (McKinsey & Co. 1966, 1967), these procedures were riven with inefficiencies, notably wasted time. Due to the irregularity of the system as a whole Gunston perceived that the workforce was essentially under-utilised, implying that the 'idleness' of the labour force could be replaced by the efficiencies of technology. He stated, for example, that due to the non-integrated nature of freight transport prior to containerisation, 75% of the North Atlantic run was spent in port rather than at sail (Gunston 1968, 59). The overarching criticisms, then, were unnecessary worker effort; inefficient labour; and wasted time. All of which could – seemingly – be resolved by the potential efficiencies of containerisation.

However, the rhetorical surety of the McKinsey & Co., reports disguises the sophistication of localised ordering procedures. It is clear from the various techniques and practices of break-bulk cargo handling and hold planning that forms of spatio-temporal ordering were indeed present. The modes of ordering described above point to practices that were developed as an ongoing counter to the spatial inconsistencies of specific cargo forms. In particular, the role of dunnage as a stabilising force was concerned with what I term the shaping of space. Here, I refer to the active attempts to regularise the shape of the load in order to create greater spatial efficiency in the hold. On a broader level, the question of shaping per se is an important one in relation to the politics of spatial ordering, for it implies the organisation of space through a regime of predetermined planning: organisation is the very foundation of 'spatial production' in the Modern era (Easterling 1999b, 3; also see Law 1994). This notion of the shaping of action bears close affinity to the idea of 'shape-determining systems' developed by the design thinker Pye (1964). For Pye there is a distinction between skilled systems that depend upon a form of constrained awareness of specific, localised qualities and specificities, and shape-determining systems that exhibit a standardised approach to the same situation. Pye's example is a bread-slicing machine (1964, 53). If bread is being cut by hand he argues the act of cutting is dependent on the interplay of factors such as the type of bread, the strength of the person, the quality of the knife. This is said to demonstrate a form of skill: 'In other words the intended result is arrived at as a rule only by way of a series of intermediate results' (Pye 1964, 54). If bread is cut by machine there is a different form of constraint, where the outcome is *predetermined*: the specific, localised qualities are reduced to a series of mechanical actions that are uniform by nature. They are continuous as opposed to intermediate phases. Such systems refer to forms of mechanical organisation, so that the actions of a component are determined or *shaped* by their arrangement within a wider system of control. Critically then, a skilled system of shaping is a localised one that is dependent on awareness of the intermediate outcomes of each stage, in contrast to a determining system that is mechanistic and generalised. Whilst both skilled systems and shape-determining systems are concerned with a form of shaping and thus organisation, it is the means of carrying this out that distinguishes them.

In the context of traditional cargo handling we can see an analogous relationship with the shaping of skilled systems. Rather than a universalised or standardised approach (as we will see with containerisation), the nature of cargo handling and packing was decidedly *localised*. That is, the spatial organisation of the hold was governed by the specific, singularised characteristics of the individual cargoes on any given vessel. The methods of shaping space, such as the use of dunnage, were part of a skill-based system of shaping exhibited through the knowledge and localised experience of the various 'nooks and crannies' of the hold (Goldblatt and Hagel 1963, 30). This notion of localised shaping resonates with Michel Serres' argument that the local is concerned with non-universalised practices that stand in contradistinction to a system of replication and repetition (Serres with Latour 1995, 91–92). Although Serres' focus is methodological the translation of this to the context of cargo packing holds firm. The traditional ordering and shaping of holds was a form of constraint that relied upon the situational qualities of the ships' hold and the nature of the cargo. Whilst there were clear modes of replication in the form of specific techniques and practices these were not mechanistic in the sense outlined by Pye.

Nascent Packaged Efficiencies: Pallets, Non-standardised Containers and the Move to Unitisation

We saw above that prior to intermodal containerisation the system of cargo handling and stowage was clearly based on an organisational logic, albeit a localised, skill-based system. As will be outlined later in this paper the development of intermodal containerisation can be seen to follow this logic of shaping by extending the skill-based modes of spatial ordering into a universal system of mechanistic shape-determination, or what I term *packaged efficiency*. However, there was a clear period of genealogical transition between the two phases, as opposed to a paradigm shift (Levinson 2006).

Up until the wide-scale implementation of intermodal containerisation in the 1970s, early forms of unitised cargo shape and mechanised cargo handling were present. Perhaps, the most notable early example of regularisation of cargo shape was the use of wooden pallets. Whilst still in evidence today in some commercial ports (House 2005, 28), pallets prefigured the uniformity and formal regularity of the intermodal shipping container. They did so by homogenising the diversity of individual break-bulk cargo shape through the loading of cargo onto a pallet. This apparently simple piece of mundane design (Heskett 2002; Michael 2003, 128; Molotch 2003; Shove et al. 2007) was to prove critical in relation to the speed of loading and unloading. Rather than the need to load and unload cargo at each interface (from ship to dockside forklift truck; from forklift to warehouse; and from warehouse to lorry) according to one advert from 1967 the pallet 'takes goods right from shed to ship without reloading – saving man-hours, cutting costs!' (PLA Monthly 1967, xxvi). This partial regularisation of break-bulk cargoes provided quicker turnaround times, teamed with greater labour productivity and simplification of the dockside in terms of limits on congestion (McKinsey & Co. 1967, 75).

Decisively for my positioning of unitisation within the wider genealogy of cargo handling, the McKinsey report of 1967 notes how palletisation represented an *intermediate* stage in the move toward a full-scale system of containerisation (McKinsey & Co. 1967, 74). The humble wooden pallet predated many of the central features of the intermodal, standardised container, as we know it today. Significantly, the questions of speed and improved labour productivity instantiated by the pallet highlight the creeping regularisation and mechanisation of worker skill, as suggested by Pye's shape-determining systems. The supposed reduction in discharge and loading rates came from the scaling-up of unit size that the pallet offered, but this resulted in less contact between individual workers and loose cargo. Indeed, the removal of labour was deemed to be a central facet of full automation (McKinsey & Co. 1967, 4). However, whilst the pallet was said to improve spatial efficiency through the unitisation of shape, thus speeding up the process of loading and discharge, it did not fully optimise it in the way that containerisation would. As the '67 report stated, 'up to eight pallets would be handled at each interchange, as opposed to one 20-ft. container' (McKinsey & Co. 1967, 75). Again, this is mapped onto the resultant reduction in labour costs: '7 tons per man hour [with pallets] to over 50 tons per man hour [with fully enclosed containers]' (McKinsey & Co. 1967, 75). Further limitations of pallets were also outlined. They did not offer the potential to unitise larger forms of cargo, being more suited to smaller scale domestic products. Once more, the economies of scale and shape emerge. If pallets increased in scale then the greater homogenisation of cargo would be possible. Finally, the later McKinsey report noted that the full, unrestricted movement of cargo on a global basis was limited due to the inability to securely seal pallets and protect from damage and theft (McKinsey & Co. 1967, 74). Through these various aspects we can appreciate the remit of unitisation more widely: to implement the most efficient process of transfer by overcoming the perceived limitations of labour, and the spatio-temporal as well as economic inefficiencies of shape and size.

The process of regularisation that the pallet began to highlight was extended further by the unitisation of wooden packing crates – a seeming extension of the cubic form of the pallet. More significantly, still this was seen with the design of fully sealed *non*-standardised containers, which at first appear almost identical in form to the later standardised intermodal containers, although they differed from each other in size, shape and fixings (Huntington 1964).[2] The idea of transporting freight in some form of sealed container had been evident since the late nineteenth Century when British and French railways used wooden boxes on flatbed rail wagons (Owen 1962; "Uniform Containerization of Freight" 1969; Levinson 2006, 29). Although in a different socio-political context there is an attendant legacy of non-standardised containers in military logistics, namely the Conex box. First introduced in the early

1950s these five-ton steel containers were used during the Korean War to ship soldiers' belongings, but by the time of the Vietnam War the Conex boxes were to play a decisive logistical role in the transportation of personal belongings, equipment and weapons (Levinson 2006, 174).[3]

These various designs of non-standardised, sealed containers smoothed-out certain inconsistencies of heterogeneous cargo shape, rendering the cargo 'unseen and untouched' (Goldblatt and Hagel 1963, 99). They multiplied the benefits of the pallet by extending the homogeneity of cargo shape beyond the base. Echoing the wider aesthetic of Modernist uniformity (Banham 1967; Sekula, 1995) their cubic shape regularised the system even more by enabling cargo to be stacked on top of one another both in ships' holds and portside, thus alleviating some of the spatial inefficiencies and inconsistencies seen with the stowing of irregularly shaped cargoes such as barrels. In the case of transporting automobiles, for example, the diverse shape of individual ones was overcome through the use of a tubular metal armature to regularise the shape (Goldblatt and Hagel 1963, 92). Overall, the potential of unitised container cargo was highlighted by Owen, when he stated:

> Most types of liquids and solids may someday be moved in sealed containers interchangeable among road, rail, air, and marine transport. Advantages would include reduction in damage and loss in the time and cost of loading and unloading. Containers may prove to be the catalyst that integrates the various components of the transport sector which are now being independently planned, financed, and operated. (Owen 1962, 410)

The regularised design of these embryonic containers (or 'vans' as they are termed) allowed time to be saved through the increase in loading and discharge speed, added to which 'dunnage [did] not have to be loaded and unloaded in the hold to make walls around cargo' (Huntington 1964, 38). Sealed containers protected cargo from damage (a common problem with the use of stevedore hooks for example), and significantly alleviated problems of petty theft by the workforce (see Huntington 1964, 76; Mars 1983, 183; House 2005, 28).[4] Fundamentally for my argument the sealed nature of the containers coupled with their cubic uniformity is suggestive of a nascent form of packaged efficiency. *Here then I situate the first reading of packaged efficiency – that of an enclosed spatial unit that affords uniformity of shape.* Rather than the heterogeneity of individual entities associated with break-bulk cargo the homogenisation of cargo shape treated space in a different manner by scaling-up individual items into the packaged efficiencies of larger-scale sealed cubic boxes. The rubric of these early containers was close to the recommendations in the McKinsey & Co., reports (1966, 3; 1967, 16) that break-bulk cargo should be treated in a similar manner to bulk cargoes such as oil, grain or sugar. With the latter the treatment of this in a bulk manner had been in place since 1942 (Goldblatt and Hagel 1963, 36). The crucial factor with oil as a commodity was that it was viewed as a form of 'homogenous standardized product' (McKinsey & Co. 1967, 5). The specific materialities of this bulk cargo were eliminated, and it was treated as a unified form. In the case of containerisation this is precisely where we see the logic of homogeneity being applied across all cargoes. Added to this the skilled-based shaping and stowage by stevedores on board vessels discussed earlier in relation to Pye's (1964) work is denuded in favour a more mechanistic, and thus shape-determining, approach to the treatment of cargo. In effect we see a tentative reduction in skilled labour in favour of a controlled, systematised approach to cargo handling.

Whilst this nascent form of packaged efficiency was evident in the unitisation of break-bulk cargo this was not the case with regard to the *mobilisation* of early containers. The attendant *system* of mobility was critically important in terms of automating the movement of the containers in ports, on ships and beyond. However, although the use of foot sacking could be seen as late as 1952 ("Stevedore Foot Sacking" 1952), forms of mechanised cargo handling were present in the late nineteenth century, with portside hydraulic cranes evident in photographs from 1888 ("Untitled Photograph" 1888; also see Zimmer 1905). An advertisement in *PLA Monthly* from 1929 (PLA Monthly 1929, xii) depicted the use of cargo handling equipment such as pneumatic ship intake and discharging plant. With the case being made that such technologies facilitated ever-faster modes of discharge: the sales pitch highlights handling speeds of grain intake of '45 tons per hour' and grain discharge of '200 tons per hour' (PLA Monthly 1929, xii). On board traditional non-containerised cargo vessels the mechanisation of labour took many forms, from the stevedore's hook as an extension of the hand (Goldblatt and Hagel 1963, 13), through the mobility of four-wheel trailers which had been in use since the turn of the nineteenth Century (Goldblatt and Hagel 1963, 27), to the most widely utilised form of mechanisation: the cargo booms utilised on ships (Ford 1950, 66–67). The cargo-boom method, for example, was a system whereby cargo was loaded onto the ship using the ship's winch. In this case the boom was used to drag cargo up a skid ramp from the dockside, where it was then positioned at the edge of the hatch, before being lowered into the hold.

However, even with the move towards unitised cargo and the mechanisation of handling methods, the systems in place were still localised and non-integrated, whereby different methods and procedures existed in parallel. They were fragmented. Pallets still had to be moved from ship to shore, then stored in warehouses before being loaded onto lorries or the rail network. Whilst the development of the early sealed containers offered some solution to this problem, the coupling of the container with the various transports interfaces on board ships and portside was still rather rudimentary. This fragmented nature meant that even standard-sized containers (which were beginning to emerge from 1966 onwards (Levinson 2006, 137)) were often loaded onto truck trailers that were not designed to carry such items. Images from this period (Goldblatt and Hagel 1963, 104) testify to still-present forms of improvisation, with the advantages of containers (be it standardised or not at this point) being outweighed by the need to lash the container onto a non-standardised truck with rudimentary roping. So although the homogeneity of container design echoed the *unified* materiality of bulk cargoes the mobility of the container did not. At the time of the McKinsey reports in the mid-late 1960s a fully implemented – or shape-determining – system was not apparent. Above all then the notion of mechanised cargo handling was thwarted by a lack: that of systemic compatibility, completeness or the *totalised* packaging of efficiency. As the next section will discuss, the spatial, temporal and above all economic imperatives for such a totalised package came with intermodalism, where systemic compatibility and completeness through the development of fully standardised *infrastructure* was the wider goal of containerisation.

Totalised Packaged Efficiency: Containerisation, Standardisation and Delegation

As articulated earlier in this paper the standardised shipping container was seen to offer a comprehensive set of benefits to the shipping and freight transport industries, notably in terms of increased productivity and cost reductions. The economic imperatives seemed clear. The uniformity of shape and size

> provided maximum internal space, maximum volume given the maximum width dictated by the physical and legal limits of road and rail traffic, and minimum waste of space in loading containers aboard ship or parking them ashore, both horizontally and vertically. (Broeze 2002, 12)

This maximisation of spatial efficiency (or the packaging of space that I outlined above) was decisive in terms of the economic benefits. Further to this 'the use of such standardised units [created] an effective multi-modal sea-and-land system with door-to-door transport from producer to consumer' (Broeze 2002, 9). The critical importance of this object lies with its intermodality: that is, its ability to transcend the divergences between land and sea through the development of an internationally recognised and *standardised infrastructure* (see Martin, 2013). Where the regularisation of container shape was one facet of packaged efficiency in this section I posit a second reading of the packaging of space – the role of standardised universal infrastructure as a wider *system* of packaged efficiency, i.e. one akin to Pye's notion of a generalised determining system as opposed to a localised one.

Before doing so it is necessary to offer a partial history of containerisation, for the narrative highlights the rather turbulent attempts to institute a global system of trade movement, and thus the amount of background work that goes into the development of technological systems (Graham and Thrift 2007). As we saw, what exemplified the early examples of containerised cargo was the lack of an integrated or packaged system of control over their movements. According to Broeze (2002, 9) the key factor that would afford integration was the *standardisation* of the design of the container so that a *globally* recognised design could be developed. Without the development of a fully interchangeable container design along with the attendant infrastructure the economic imperative of containerisation as a totalised system was said to be limited.

It is widely acknowledged that the individual responsible for the early development of the shipping container was the US truck operator Malcom McLean (Levinson 2006, 36–53). In 1953 McLean proposed transporting truck trailers on ships rather than the congested highways of the US's east coast. Critically, the rationale for this was to overcome road congestion by consolidating opposing transport systems: at this time the road haulage and shipping industries were entirely separate (Levinson 2006, 43). There were certain limitations to the idea, including the inefficiency of transporting truck trailers with their wheels attached. Like the stowage of barrels, the irregular shape of the truck trailers meant that space was wasted under the trailer chassis. If the trailer wheels were removed the spatial wastage would be eliminated, and perhaps more fundamentally, like non-standard containers it meant that 'trailer bodies could be stacked' (Levinson 2006, 47). This was an important development, but one that, in essence, did not differ markedly from the earlier processes of unitisation outlined above. The fundamental change was the recognition that the system *as-a-whole* needed to be reconfigured and reorganised to enable the demounted trailer bodies to be moved across a range of transport networks: ships, trains and lorries (Broeze 2002, 31–32; Levinson 2006, 53). Rather than separate and competing entities the consolidation of the various freight transport interests through the development of the intermodal container required a wider programme of packaged efficiencies (Levinson 2006, 51).

McLean commissioned the container engineer Keith Tantlinger to design a new aluminium container, and to reconfigure a decommissioned tanker vessel, the *Ideal-X*, to accommodate the new containers, with no other cargo being stowed. Where previously the non-standardised containers were lifted via shipboard winches or dockside cranes using rope, McLean opted to refit two existing cranes, moving them to the ports where the first container-ship journey would be made. A further piece of equipment, the container spreader bar, was developed by Tantlinger. This enabled the container to be lifted without the need for dockworkers to attach roping. The design of the spreader bar meant that 'once the box had been lifted and moved, another flip of the switch would disengage the hooks, without a worker on the ground touching the container' (Levinson 2006, 51).

The date of 26 April 1956 is significant – it was the first sailing of the *Ideal-X* from Newark to Houston. This was important as it prefigured the momentous shifts that would occur not only throughout the shipping industry, but also across the entire transport infrastructure. Ultimately, the success of McLean's container lay with the realisation that the complete system of transportation had to be reconfigured. This relied upon the standardisation and regularisation of procedures and materials across the industry, thus ensuring systemic compatibility.

This is precisely the area that the McKinsey reports argued needed be fully recognised. They suggested that without a standardised design 'nonstandard containers by themselves are just another form of unitisation similar to pallets' (McKinsey & Co. 1967, 6). The now widely recognised standard sizes of the 'twenty-foot equivalent' (or TEU) shipping container (8 feet wide, 8 feet high, and 10, 20, 30 or 40 feet long) were only fully agreed as late as 1970 by International Organization for Standardization (ISO) (Levinson 2006, 148). Although the initial sizes were agreed in 1961 (Egyedi 2001, 49; Levinson 2006, 137) it was only after 1966 that various interested parties in the shipping industry began to compromise. Vital to structural integration was the standardised nature of infrastructure, enabling the coupling of the container across a variety of nodes. Everything had to be effectively designed from scratch. These included numerous significant technical developments, such as container-cell ships to accommodate containers in specially designed cell bays on the vessels (Goldblatt and Hagel 1963, 104–107; Pinder and Slack 2004, 3); the redesign of road haulage vehicles and railway rolling stock; the design of container handling vehicles in ports; the construction of large-scale dockside gantry cranes ("Container Transporter Crane" 1970); the design of the spreader bars (Levinson 2006, 51); and the design of the container corner fittings (Martin, 2013).

In this final section I turn to the importance of standardisation in relation to the wider conceptualisation of packaged efficiencies. In particular, standardisation embodies the shift away from the packaging of localised skill-based processes of shaping into a shape-determining system of predetermined action. Critically, my rendering of a system of packaged efficiency directly refers to the construction of a defined spatial, temporal, technological and networked apparatus that attempts to facilitate stable forms of connection and linkages with other facets of the system

(Law 2000, 4). Obviously this definition could be afforded to a range of applications. Crucially, in the context of cargo handling and the broader system of containerisation it is the maintenance of mobility through standardised and predetermined forms of linkage that are critical.

The final agreement over the standardised design and dimensions of the ISO intermodal container points to the way in which the regularisation of design ensured compatibility across the various transport networks (Gunston 1968, 59). Standardisation in the manufacturing process had been in existence since the impact of mechanisation and notably in relation to the use of interchangeable machine parts (see Higgins and Hallström 2007, 691; Nye, 2013, 27–28), as well as in the manufacture of firearms (DeLanda 1991, 31). In the case of manufacturing, the standardisation and interchangeability of parts was intended to eradicate limitations in the flows of parts within a production line. Or more broadly,

> The development of technology involves not just the reduction of blockages through the production of technical standards and other mechanisms, but the development of ways of circumventing or reconfiguring existing impediments and ways of establishing new ones. (Barry 2001, 18)

This outline foregrounds the intricate relationship between the role of standard procedures, standardised materials and the diminution in impediments to specific circumstances (in our case the packaging of goods in ships' holds and the flow of goods across various transport platforms). Added to this, standardisation creates a boundary that protects the system of flow itself i.e. my notion of the system of packaged efficiency.

Barry identifies how standards are concerned with overcoming systemic division through coordination, organisational logic, and systemic completeness. Higgins and Larner (2010, 3–4) locate technological standards within a field 'which makes social domains knowable and governable'. Such practices and procedures of governance are intrinsic to the development of global parameters, so that 'at a distance' control over specific global processes can be actioned. This is where we see a critical distinction between localised, skill-based parameters and those that are generalised and universalised. Effectively, standards and standardisation overcome the systemic 'messiness' of localised knowledge that Turnbull (1993, 317) describes, in that it lacks the ability to be transmitted globally.

In this context the system of containerisation is premised on the coordination and transmission of *connection,* i.e. the relations between all of the constituent elements of the system. As the material devices such as the spreader bars and container corner fitting demonstrate, there has to be a stabilisation of relations, whereby the linkage between the container and the various modes of transport is guaranteed through universally recognised design, as opposed to localised tactics of traditional cargo handling. More pointedly, there is a stabilisation of change, where the mobility of the container is guaranteed as it moves across different transport platforms. There has to be a guaranteed *fit* (Star 1991). The notion of guarantee is an important one in the work on standardisation. Bowker and Star (2000, 13) note that standards are 'any set of agreed-upon rules for the production of (textual or material) objects', whilst Brunsson and Jacobsson (2002, 15) confirm this by defining standardisation as the *implementation* of agreed-upon rules. The development of the intermodal container exemplifies this: only with the collective agreement on the standardised sizes could there be the guarantee of its regularity of design and its attendant ability to interact with other material objects, or indeed other standards (see Fabbe-Costes, Jahre, and Rouquet 2006). Through such structures of universal agreement standardisation embodies the entrenchment of standards so that the various standardised components of a system become 'crystallised' (Egyedi 2001, 41), or what I term *packaged* into a shape-determining system. That is, they are congealed into a working package, where a boundary is created between those objects that have been agreed-upon, and those that have not.[5] Through this process of entrenchment one sees the selection of specific technologies that are designed to be universally compatible. For Bowker and Star (2000, 13) the notion of compatibility highlights an important spatial feature of standards: that is, how 'a standard spans more than one community of practice (or site of activity)'. This is achieved across both space and time: i.e. standardised objects are able to operate across distance, but also to sustain compatibility and reliability over time. Both aspects of this relationship embody the crux of the intermodal container: it is built on its ability to interchange through the infrastructure of containerisation, and equally the embedded or entrenched nature of the infrastructure guarantees the *ongoing* ability to interchange.

This argument concerning standardisation and inter-changeability speaks to the importance of stabilising interconnection through the predetermination allied to shape-determining systems. Latour (1992) reads the various processes of stabilisation via the concept of delegation. Delegation is a process whereby human effort is delegated to socio-technical machines, be that a washing machine, television remote control (Michael 2000) or door hinge in Latour's example.[6] Following in the long analytical tradition of studies on automation (Gideion, 1948; Ellul 1964) an object may be said to 'displace, translate, delegate, or shift' (Latour 1992, 229) its function from one of major effort (opening a heavy door) to a minor one (displacing this into the light push of the door). Here, we can posit a direct correlation with both Pye's work on skilled and shape-determining systems, and my rendering of packaged efficiency. The opening of a door, or the traditional means of packing a ships' hold requires an intermediate set of relationships with the specific circumstances, whereas the door hinge or container spreader bar delegates the localised affects into a mechanical form of continuity. Through the design of the apparently simple devices of the container corner fitting, the spreader bar, or container guide posts and runners onboard container ships, the previous work, effort and time expended on conjoining cargoes and vehicle (through the lashing of ropes) is built into the device: it is *delegated* to it. The expenditure of effort in lashing ropes around non-standardised containers or utilising dunnage to protect cargo is seemingly replaced by the 'dream of efficient action' (Latour 1992, 235) and packaged in the technical device itself. That is to say the purpose of these devices (to link container and vehicle) replaces the previous job of the dockworker (Goldblatt and Hagel 1963, 106). Likewise it could be argued that the technical know-how and skill-based knowledge of stowage that stevedores previously utilised becomes packaged in the system of containerisation i.e. the shape-determining system of automation and unitisation alleviates the need for the skill-based stowage of cargo because these actions are packaged into the system of containerisation.

Conclusions

Delegation is an inherently political process (see Adorno 1996, 40), and in the context of the shipping industry a devastating one in relation to the destruction of traditional maritime communities (Goldblatt and Hagel 1963; Bonacich and Wilson 2008, 15–22), forms of deindustrialisation (Harvey 2010, 16), whilst also embodying the shift in the nature of maritime labour from community embeddedness to precarity (Sekula 2000). As such we can see how containerisation speaks to the political geographies of labour relations and the impact of technology more broadly. It is a decisive example of the automation of mobility on a global scale and one that links across a variety of similar examples from the Modern industrial era (Taylor 1911; Giedion 1948; Hounshell 1984, 249–253; Bahnisch 2000; Cresswell 2006, 95–115). However, we have to be critically cognisant of the fact that even with the delegation to technical objects and the packaging of skill-based processes into shape-determining systems, the 'dream of efficient action' referred to by Latour is, in effect, an idealised image of efficiency. For the numerous maritime accidents (Cresswell and Martin 2012), geopolitical tensions (Middleton 2008) and global economic effects (Rosenthal 2010) attest to the incredible amount of background work that goes into maintaining the package of efficiency (Graham and Thrift 2007). And in part this has been one of the underlying intentions of this paper: to consider how the conceptualisation of packaged efficiency will facilitate a wider appreciation of how and why specific processes, spaces, material forms, systems and processes are packaged.

Returning to the opening example of the increasing size of container ships it is clear that the economies of scale identified in the McKinsey & Co., reports in the 1960s are still intrinsic to the contemporary geographies of ships. The story also highlights how ships are only one actor in the wider package of containerisation. As we have seen, the scaling-up of container ships to accommodate 18,000 TEU's is part of a much longer genealogy of attempts to increase the efficiencies of commodity mobilities, particularly with regard to speed and cost effectiveness.

To be sure, spatio-economic parameters pervade the logistical mindset: Cowen confirms this by pointing out that logistics management (as part of the wider armature of containerisation) is tasked with 'annihilating minutes or even seconds from transactions along supply chains' (Cowen 2010a, 602). As stated at the outset to this paper, the genealogy of the spatial ordering of ships' holds prefigures the wider contemporary geographies of logistical power (Neilson 2012). To this end the technological development of containerisation, including the profound effects on the geographies of ships and the shipping industry, highlights its centrality to the spatial logic of contemporary capitalism (Harvey 2010).

Whilst the considerable cultural, social, geographical, economic and labour changes instituted by the global acceptance of containerisation have been profound, it is clear from the arguments raised here that there is a distinct lineage in place. One where the increasing will to create the most efficient movement of cargo can be read as emanating out of the previous spatial organisation of cargo holds. My key argument was that the packaging of efficiency is not solely a product of the mid-late twentieth century embrace of containerisation but rather is imbued within a genealogy of spatio-temporal organisation, seen in the first section of the paper with the ordering of the cargo hold to provide a tight stow. As this section on pre-containerisation suggested there was recognition of the inconsistencies of cargo shape and form, necessitating the need for strategies such as dunnage to regularise the load, both for the purposes of stowage and to stabilise the load during sail.

One of the key debates throughout this paper has been the importance of regularisation, homogenisation and unitisation to spatial theory. Where the spatial inconsistencies of break-bulk cargo were evident, the early processes of unitisation in the form of pallets and packing crates represented attempts to formalise and regularise inconsistencies. This was then developed further by the introduction of fully sealed containers. The benefits of such material strategies were evident to see, including the protection of cargo from damage or theft; the cubic efficiency of the container; increased speed of loading and discharge; reduction in labour costs; and the ability to interchange between different forms of transport. This final impact highlights the obvious kinship between these early sealed containers and the later fully standardised intermodal units. Using David Pye's notions of the difference between skill-based and shape-determining systems it is possible to suggest that non-standardised containers (and perhaps pallets before them) represented a form of local knowledge and stability, whereby the uniformity and regularisation of cargo enabled partial forms of inter-changeability. Thus, they exhibited what I termed nascent packaged efficiencies: a spatial uniformity that facilitated increased organisation of ships' holds. However, my argument was that they lacked *global* stability and inter-changeability. So, although the efficiencies of the material packaging of early containers along with forms of partial automation and mechanisation posited the increasing will to alleviate 'wasted' time and manpower, the relationship between handling procedures and cargo items was still relatively fragmented.

As suggested in the section on totalised packaged efficiencies, what epitomised the spatio-economic ideology of intermodal containerisation was the role of packaged *completeness* i.e. systemic control across the entire freight transport infrastructure (road, rail and maritime) and leading into the spatiality of contemporary supply chains (Hughes and Reimer 2004). By identifying this approach the intention was to posit the potential link between the overarching ideology of spatial modes of ordering and the ethos of containerisation: that is, how the standardisation of space-time delineates the global mobility of containerisation as a single systemic package and its attendant maintenance, development and securitisation.

Notes

1. It should be noted that even with the move towards fully standardised containers the nature of packing the containers themselves means that dunnage is still utilised, albeit in the form of air-filled packaging materials. As discussed at the end of this paper, whilst the notion of packaged efficiency is used to highlight standardised practices there are still improvised practices and legacies of earlier cargo-handling procedures.
2. Early examples of containerised cargo shipments include the Link-Line service between Liverpool-Belfast, started in January 1959. This service used 12-ton capacity aluminium containers, but these were non-standard in design and used rounded top edges ("Link-Line Service Liverpool-belfast" 1959). Levinson (2006, 31) notes that a similar service was in operation in Denmark in 1951, and the Transportainer was developed by the Pittsburgh-based Dravo Corporation in 1954.
3. The phasing-out of Conex boxes for specifically military purposes in 1968 coincided with the growing dominance of the commercial shipping container.
4. In an earlier example from the eighteenth century the importation of loose tobacco from Maryland and Virginia was outlawed, and the stipulation made that all tobacco be imported in casks, chests or hogshead cases in order to reduce the potential for the smuggling of other goods inside the bundles of tobacco (Rive 1929, 558).
5. This is not to suggest that non-agreed-upon objects always lie outside of such standardised systems, rather that certain levels of improvisation are required in this case.
6. Like Pye before him Latour notes that delegation does not solely move from human to non-human. Instead it can be a process of delegation to a 'more durable' actor, be they human or non-human (Latour 1992, 256 n.6).

References

Adorno, T. 1996. *Minima Moralia*. London: Verso.

Airriess, C. A. 2001. "Regional Production, Information-communication Technology, and the Developmental State: The Rise of Singapore as a Global Container Hub." *Geoforum* 32: 235–254.

Allen, W. B. 1997. "The Logistics Revolution and Transportation." *The ANNALS of the American Academy of Political and Social Science* 553: 106–116.

Bahnisch, M. 2000. "Embodied Work, Divided Labour: Subjectivity and the Scientific Management of the Body in Frederick W. Taylor's 1907 'Lecture on Management'." *Body & Society* 6 (1): 51–68.

Banham, R. 1967. "Flatscape with Containers." *New Society*: 231–232. 17 August.

Barry, A. 2001. *Political Machines: Governing a Technological Society*. London: Continuum.

Bohlman, M. 2001. "ISO's Container Standards are Nothing but Good News: Containers Standards Help to Remove Technical Barriers to Trade." *ISO Bulletin*: 12–15. September.

Bonacich, E., and J. B. Wilson. 2008. *Getting the Goods: Ports, Labour, and the Logistics Revolution*. Ithaca, NY: Cornell University Press.

Bowker, G. C., and S. L. Star. 2000. *Sorting Things Out: Classification and its Consequences*. Cambridge, MA: MIT Press.

Branch, A. E. 2007. *Elements of Shipping*. 8th ed. Abingdon: Routledge.

Bratton, B. H.. 2006. "Introduction: Logistics of Habitable Circulation." In *Speed and Politics*, edited by P. Virilio, 7–25. New York: Semiotext(e).

Broeze, F. 2002. *The Globalization of the Oceans: Containerisation from the 1950s to the Present*. St. John's: International Maritime Economic History Association.

Brunsson, N., and B. Jacobsson. 2002. *A World of Standards*. Oxford: Oxford University Press.

Buchloh, B., D. Harvey, and A. Sekula. 2011. "Forgotten Spaces: Discussion Platform with Benjamin Buchloh, David Harvey, and Allan Sekula, at a Screening of 'The Forgotten Space' at The Cooper Union, May 2011." Accessed January 25, 2012. http://www.afterall.org/online/material-resistance-allan-sekula-s-forgotten-space

Cargoes. 1940. *Directed by Humphrey Jennings*. London: GPO.

Cidell, J. 2012. "Flows and Pauses in the Urban Logistics Landscape: The Municipal Regulation of Shipping Container Mobilities." *Mobilities* 7 (2): 233–245.

"Clan MacDougall [Neg. P.39542]." 1938. Box 90.1, National Maritime Museum Photographic Archive, London.

"Container Transporter Crane." 1970. *Design* 258 (June): 66–69.

Corry, P., and E. Kozan. 2008. "Optimised Loading Patterns for Intermodal Trains." *OR Spectrum* 30: 721–750.

Cowen, D. 2010a. "A Geography of Logistics: Market Authority and the Security of Supply Chains." *Annals of the Association of American Geographers* 100 (3): 600–620.

Cowen, D. 2010b. "Containing Insecurity: Logistics Space, U.S. Port Cities, and the 'War on Terror'." In *Disrupted Cities: When Infrastructure Fails*, edited by S. Graham, 69–83. Abingdon: Routledge.

Cresswell, T. 2006. *On the Move: Mobility in the Modern World*. London: Routledge.

Cresswell, T., and C. Martin. 2012. "On Turbulence: Entanglements of Disorder and Order on a Devon Beach." *Tijdschrift voor economische en sociale geografie* 103 (5): 516–529.

Cudahy, B. 2006. *Box Boats: How Container Ships Changed the World*. New York: Fordham University Press.

DeLanda, M. 1991. *War in the Age of Intelligent Machines*. New York: Zone Books.

Dicken, P. 2011. *Global Shift: Mapping the Changing Contours of the World Economy*. London: Sage.

Easterling, K. 1999a. "Interchange and Container: The New Orgman." *Perspecta* 30: 112–121.

Easterling, K. 1999b. *Organization Space: Landscapes, Highways, and Houses in America*. Cambridge, MA: MIT Press.

Egyedi, T. 2001. "Infrastructure Flexibility Created by Standardized Gateways: The Cases of XML and the ISO Container." *Knowledge, Technology, & Policy* 14 (3): 41–54.

Ellul, J. 1964. *The Technological Society*. New York: Vintage.

Fabbe-Costes, N., M. Jahre, and A. Rouquet. 2006. "Interacting Standards: A Basic Element in Logistics Networks." *International Journal of Physical Distribution & Logistics Management* 36 (2): 93–111. http://proquest.umi.com/pqdweb?did=1028907171&Fmt=3&clientId=52553&RQT=309&VName=PQD

Ford, A. G. 1950. *Handling and Stowage of Cargo*. Scranton, PA: International Textbook.

Forrester, J. W. 1958. "Industrial Dynamics: A Major Breakthrough for Decision Makers." *Harvard Business Review* 38: 37–66. July–August.
Foucault, M. 1977. "Nietzsche, Genealogy, History." In *Language, Counter-memory, Practice: Selected Essays and Interviews*, edited by D.F. Bouchard, 139–164. Ithaca, NY: Cornell University Press.
Giedion, S. 1948. *Mechanization Takes Command: A Contribution to Anonymous History*. New York: W.W. Norton and Co.
Gilroy, P. 1993. *The Black Atlantic: Modernity and Double Consciousness*. London: Verso.
Goldblatt, L., and O. Hagel. 1963. *Men and Machines: A Story about Longshoring on the West Coast Waterfront*. San Francisco, CA: International Londshoremen's and Warehousmen's Union.
Gomes, R., and J. T. Mentzer. 1988. "A Systems Approach to the Investigation of Just-in-time." *Journal of Business Logistics* 9 (2): 71–88.
Graham, S., and N. Thrift. 2007. "Out of Order: Understanding Repair and Maintenance." *Theory, Culture & Society* 24 (3): 1–25.
Gunston, B. 1968. "Moving the Goods." *Design* 234: 58–62. June.
Gutting, G. 1990. "Foucault's Genealogical Method." *Midwest Studies in Philosophy* 15 (1): 327–343.
Harvey, D. 2010. *The Enigma of Capital and the Crises of Capitalism*. London: Profile Books.
Hasty, W., and K. Peters. 2012. "The Ship in Geography and the Geography of Ships." *Geography Compass* 6 (7): 660–676.
Heskett, J. 2002. *Toothpicks and Logos: Design in Everyday Life*. Oxford: Oxford University Press.
Higgins, W., and K. T. Hallström. 2007. "Standardization, Globalization and Rationalities of Government." *Organization* 14 (5): 685–704.
Higgins, W., and W. Larner. 2010. "Standards and Standardization as a Social Scientific Problem." In *Calculating the Social: Standards and the Reconfiguration of Governing*, edited by V. Higgins and W. Larner, 1–17. Basingstoke: Palgrave MacMillan.
Hounshell, D. A. 1984. *From the American System to Mass Production, 1800–1932: The Development of Manufacturing Technology in the United States*. Baltimore, MD: Johns Hopkins University Press.
House, D. J. 2005. *Cargo Handling for Maritime Operations*. Oxford: Elsevier Butterworth-Heinemann.
Hughes, A., and S. Reimer. 2004. "Introduction." In *Geographies of Commodity Chains*, edited by A. Hughes and L. Reimer, 1–16. London: Routledge.
Hunter, P. 1993. *The Magic Box: A History of Containerization*. Ottawa: ICHCA Canada.
Huntington, H. E. 1964. *Cargoes*. Garden City, NY: Doubleday and Co.
Institute of Shipping Economics and Logistics. 2011. *Shipping Statistics and Market Review* 55 (5/6): 5–52.
Jackson, G. 1983. *The History and Archaeology of Ports*. Tadworth: World's Work.
Kanngieser, A. 2013. "Tracking and Tracing: Geographies of Logistical Governance and Labouring Bodies." *Environment and Planning D: Society and Space* 31(4): 594–610.
Latour, B. 1992. "Where are the Missing Masses? The Sociology of a Few Mundane Artifacts." In *Shaping Technology/Building Society: Studies in Sociotechnical Change*, edited by W. E. Bijker and J. Law, 225–258. Cambridge, MA: MIT Press.
Law, J. 1994. *Organizing Modernity*. Oxford: Blackwell.
Law, J. 2000. "Objects, Spaces and Others." Accessed March 5, 2005. http://www.comp.lancs.ac.uk/sociology/papers/Law-Objects-Spaces-Others.pdf
Law, J. 2003. "On the Methods of Long Distance Control: Vessels, Navigation, and the Portuguese Route to India." Accessed January 12, 2006. http://www.comp.lancs.ac.uk/sociology/papers/Law-Methods-of-Long-Distance-Control.pdf
Levinson, M. 2006. *The Box: How the Shipping Container made the World Smaller and the World Economy Bigger*. Princeton, NJ: Princeton University Press.
"Link-Line Service Liverpool-belfast [Photograph]." 1959. Box 90.1, National Maritime Museum Photographic Archive, London.
Mars, G. 1983. *Cheats at Work: An Anthropology of Workplace Crime*. Winchester, MA: Allen & Unwin.
Martin, C. 2011. "Desperate Passage: Violent Mobilities and the Politics of Discomfort." *Journal of Transport Geography* 19: 1046–1052.
Martin, C. 2012. "Containing (Dis)order: A Cultural Geography of Distributive Space." Unpublished PhD thesis, Royal Holloway, University of London.
Martin, C. 2013. "Shipping Container Mobilities, Seamless Compatibility, and the Global Surface of Logistical Integration." *Environment and Planning A* 45 (5): 1021–1036.

McCalla, R. J. 1999. "Global Change, Local Pain: Intermodal Seaport Terminals and their Service Areas." *Journal of Transport Geography* 7 (4): 247–254.
McKinsey & Company, Inc. 1966. *Containerization – Its Trends, Significance and Implications*. London: McKinsey & Co.
McKinsey & Company, Inc. 1967. *Containerization: The Key to Low-cost Transport*. London: McKinsey & Co.
Mezzadra, S., and Neilson, B. 2013. "Extraction, Logistics, Finance: Global Crisis and the Politics of Operations." *Radical Philosophy* 178 (March/April): 8–19.
Michael, M. 2000. *Reconnecting Culture, Technology and Nature: From Society to Heterogeneity*. London: Routledge.
Michael, M. 2003. "Between the Mundane and the Exotic: Time for a different Sociotechnical Stuff." *Time & Society* 12 (1): 127–143.
Middleton, R. 2008. *Piracy in Somalia: Threatening Global Trade, Feeding Local Wars (Briefing Paper)*. London: Chatham House.
Molotch, H. 2003. *Where Stuff Comes From: How Toasters, Toilets, Cars, Computers and Many Other Things Come To Be As They Are*. New York: Routledge.
Neate, R. 2013. "Giants of the Sea Force Ports to Grow." *The Guardian*, 7 March, 33.
Neilson, B. 2012. "Five Theses on Understanding Logistics as Power." *Distinktion: Scandinavian Journal of Social Theory* 13 (3): 323–340.
Nye, D. 2013. *America's Assembly Line*. Cambridge, MA: MIT Press.
Owen, W. 1962. "Transportation and Technology." *The American Economic Review* 52: 405–413.
Parker, C. H. 2010. *Global Interactions in the Early Modern Age, 1400–1800*. Cambridge: Cambridge University Press.
Pinder, D., and B. Slack. 2004. "Contemporary Contexts for Shipping and Ports." In *Shipping and Ports in the Twenty-first Century: Globalisation, Technological Change, and the Environment*, edited by D. Pinder and B. Slack, 1–22. London: Routledge.
PLA Monthly. 1929. *PLA Monthly*, April.
PLA Monthly. 1967. *PLA Monthly*, May.
Port of London Authority. 1979. *Five Year Strategic Plan 1979–83*. London: Port of London Authority (Museum of Docklands Archive: loc. 2/2/4: PLA Development Folder c.1960s-1970s).
Pye, D. 1964. *The Nature of Design*. London: Studio Vista.
Rive, A. 1929. "A Short History of Tobacco Smuggling." *The Economic Journal/Economic History Supplement* 1 (4): 554–569.
Rosenthal, E. 2010."Slow Trip Across Sea Aids Profit and Environment." Accessed February 5, 2012. http://www.nytimes.com/2010/02/17/business/energy-environment/17speed.html
Schmeltzer, E., and R. A. Peavy. 1970. "Prospects and Problems of the Container Revolution." *The Transportation Law Journal* 2 (2): 263–299.
Sekula, A. 1996. *Fish Story*. Düsseldorf: Richter Verlag.
Sekula, A. 2000. "Freeway to China (version 2, for Liverpool)." *Public Culture* 12: 411–422.
Serres, M., with B. Latour. 1995. *Conversations on Science, Culture and Time*. Ann Arbor: University of Michigan Press.
Shaw, J., and J. Sidaway. 2010. "Making Links: On (Re)engaging with Transport and Transport Geography." *Progress in Human Geography* 35 (4): 502–520.
Shove, E., M. Watson, M. Hand, and J. Ingram. 2007. *The Design of Everyday Life*. Oxford: Berg.
Smith, A. 1989. "Gentrification and the Spatial Constitution of the State: The Restructuring of London's Docklands." *Antipode* 21: 232–260.
Star, S. L. 1991. "Power, Technology and the Phenomenology of Conventions: On Being Allergic to Onions." In *A Sociology of Monsters: Essays on Power, Technology and Domination*, edited by J. Law, 26–56. London: Routledge.
Steinberg, P. 2001. *The Social Construction of the Ocean*. Cambridge: Cambridge University Press.
"Stevedore Foot Sacking [Neg. C.6715]." 1952. Box 90.1, National Maritime Museum Photographic Archive, London.
Talley, W. K. 2000. "Ocean Container Shipping: Impacts of a Technological Improvement." *Journal of Economic Issues* 34 (4): 933–948.
Taylor, F. W. 1911. *The Principles of Scientific Management*. New York: Harper & Brothers.
Teräs, K. 2007. "Discourse and the Container Revolution in Finland in the 1960s and 1970s." In *Making Global and Local Connections: Historical Perspectives on Ports (Research in Maritime History*

No.35), edited by T. Bergholm, L. R. Fischer, and M. E. Tonizzi, 137–152. St. John's: International Maritime Economic History Association.
Tomlinson, J. 2007. *The Culture of Speed: The Coming of Immediacy*. London: Sage.
Toscano, A. 2011. "Logistics and Opposition." *Mute: Culture and Politics after the Net*. Accessed August 10. http://www.metamute.org/editorial/articles/logistics-and-opposition
Turnbull, D. 1993. "The Ad Hoc Collective Work of Building Gothic Cathedrals with Templates, String, and Geometry." *Science, Technology and Human Values* 18 (3): 315–340.
"Uniform Containerization of Freight: Early Steps in the Evolution of an Idea." 1969. *Business History Review* 43 (1): 84–87.
"Untitled Photograph. [Neg. B8635]." 1888. Box 90.1, National Maritime Museum Photographic Archive, London.
Wrigley, N. 2000. "The Globalization of Retail Capital: Themes for Economic Geography." In *The Oxford Handbook of Economic Geography*, edited by G. L. Clark, M. S. Gertler, and M. P. Feldman, 292–313. Oxford: Oxford University Press.
Zimmer, G. F. 1905. *The Mechanical Handling of Material*. London: Crosby Lockwood & Son.

Index